This book is affectionately dedicated to
my younger brother, Balwant S. Dhillon

Contents

Preface

Today, engineering systems are an important element of the world economy, and each year, a vast sum of money is spent to develop, manufacture, operate, and maintain various types of engineering systems around the globe. The reliability and usability of these systems have become more important than ever because of their increasing complexity, sophistication, and nonspecialist users. Global competition and other factors are forcing manufacturers to produce highly reliable and usable engineering systems.

This fact means that there is a definite need for reliability and usability professionals to work closely during the design of a system, as well as other phases. To achieve this goal, it is essential that they have to a certain degree an understanding of each other's discipline. At present, to the best of the author's knowledge, there is no book that covers the topics of reliability and usability within this framework. This means that, at present, to gain knowledge of each other's specialities, these professionals must study a wide range of books, reports, and articles on each of the topics in question. This approach is time-consuming and rather difficult because of the specialized nature of the material involved.

Thus, the main objective of this book is to combine these two topics into a single volume and to eliminate the need to consult many diverse sources in order to obtain basic and up-to-date information on these topics. The sources of most of the material presented are given in the reference sections in the chapters. These will be useful to readers if they desire to delve more deeply into a specific topic or area. The topics covered in this book are treated in such a manner that the reader will require no previous knowledge to understand the text. At appropriate places, this book contains examples along with their solutions, and there are numerous problems at the end of each chapter to test reader comprehension in the area in question.

This book is composed of 13 chapters. Chapter 1 presents the need for and the historical developments in reliability and usability; engineering systems reliability/usability-related facts, figures, and examples; important terms and definitions; and useful sources of information on reliability and usability. Chapter 2 reviews mathematical concepts considered useful to understand subsequent chapters in this book. Some of the topics covered in the chapter are Boolean algebra laws, probability properties, probability distributions, and useful mathematical definitions.

Chapter 3 presents introductory aspects of reliability and of human factors, directly or indirectly, concerned with usability. Chapter 4 presents a number of methods considered useful to evaluate the reliability and usability of engineering systems, including failure modes and effect analysis, the Markov method, fault tree analysis (FTA), cognitive walkthroughs, cooperative evaluation, task analysis, property checklists, expert appraisals, probability tree analysis, and cause-and-effect diagrams. Chapter 5 is devoted to robot system and medical equipment reliability. Some of the topics covered are robot failure causes and classifications, robot reliability analysis methods and models, electric and hydraulic robots' reliability analysis, medical equipment reliability improvement procedures and methods, human error in the use of medical equipment, and medical equipment maintenance–related indices.

Chapter 6 presents several important aspects of transportation system failure and oil and gas industry equipment reliability. Some of the topics covered in the chapter are mechanical failure–related aviation accidents, defects in vehicle parts and classifications of vehicle failures, rail defects and weld failures, ship failures and their causes, corrosion-related failures, and oil and gas pipeline FTA. Chapter 7 is devoted to computer system and Internet reliability and software bugs in computer systems, and covers topics such as sources of computer failures, computer-related fault categories and reliability measures, fault masking, Internet reliability–related observations, mathematical models for performing Internet reliability and availability analysis, methods for preventing programmers from introducing bugs during the software-writing process, and software error–related metrics.

Chapter 8 is devoted to power system and mining equipment reliability. The topics covered in the chapter include service performance–related indexes, loss-of-load probability, mathematical models for performing availability analysis of transmission and associated systems, reasons for improving mining equipment reliability, mining equipment reliability measures, methods for measuring winder rope degradation, and open-pit system reliability analysis. Chapter 9 presents various important aspects of usability engineering life-cycle stages and important associated areas. Some of the topics covered in the chapter are usability engineering life-cycle stages, fundamental features of design for usability, usability-related myths and factors affecting usability within organizations, usability performance measures, and usability advantages.

Chapter 10 is devoted to usability testing and covers topics such as usability-testing goals, limitations, and advantages; usability-testing elements, types of usability-related tests, usability cost–related facts and figures, usability engineering–related activities and costs, and models for estimating usability-engineering costs. Chapter 11 presents important aspects of software and web usability. Some of the topics covered in the chapter are the need to consider usability during software development, the software-usability-engineering process, steps for improving the usability of software products, software-usability-testing methods, common web design–related errors, web page design, tools for evaluating web usability, and navigation aids.

Chapter 12 is devoted to medical device usability and user errors. Topics covered in the chapter include medical device use, users, user environments, and user interfaces; the general approach for developing effective user interfaces of medical devices, designing medical devices for older users, medical devices with high incidence of user/human error, common medical device/equipment operator/user errors, and methods for performing user error–related analysis. Finally, Chapter 13 presents 15 mathematical models for systems reliability analysis and usability assurance.

This book will be useful to a wide variety of professionals, including design engineers, system engineers, reliability and usability professionals, human factors and ergonomics professionals, computer system engineers, computer-interface specialists, graduate and senior undergraduate students, researchers and instructors in the area of reliability and usability engineering, and engineers in general.

B. S. Dhillon
Ottawa, Ontario

Author

Dr. B. S. Dhillon is a professor of engineering management in the Department of Mechanical Engineering at the University of Ottawa. He has served as a chairman/director of the Mechanical Engineering Department/ Engineering Management Program for more than 10 years at the same institution. He is the founder of the probability distribution named the *Dhillon Distribution/Law/Model* by statistical researchers around the world. He has authored or co-authored more than 397 articles on reliability engineering, maintainability, safety, engineering management, and other topics. He is or has been on the editorial boards of 12 international scientific journals. In addition, Dr. Dhillon has written 46 books on various aspects of health care, engineering management, design, reliability, safety, and quality. His books are being used in more than 100 countries, and many of them are translated into languages such as German, Russian, Chinese, Arabic, and Persian.

He has served as the general chairman of two international conferences on reliability and quality control held in Los Angeles and Paris in 1987. He has also served as a consultant to various organizations and bodies and has many years of experience in the industrial sector. He has lectured in over 50 countries, including keynote addresses at international scientific conferences held in North America, Europe, Asia, and Africa. In March 2004, he was a distinguished speaker at the Conference/Workshop on Surgical Errors (sponsored by the White House Health and Safety Committee and the Pentagon) in Washington, D.C.

Professor Dhillon attended the University of Wales, where he received a BS in electrical and electronic engineering and an MS in mechanical engineering. He received a PhD in industrial engineering from the University of Windsor.

chapter one

Introduction

1.1 Background

The history of the reliability discipline may be traced back to the early 1930s, when probability concepts were applied to problems concerning electric power generation [1,2]. During World War II, Germans applied the basic reliability concepts to improve the reliability of their V1 and V2 rockets. During the period of 1945–1950, the U.S. Department of Defense conducted various studies concerning electronic equipment failure, equipment maintenance, and repair cost. As the result of these studies, in 1950, it formed an ad hoc committee on reliability, and in 1952, the committee was transformed to a permanent body: Advisory Group on the Reliability of Electronic Equipment (AGREE) [3]. Additional information on the history of the reliability discipline is available in References 4,5.

The emergence of the usability engineering field is deeply embedded in the discipline of human factors. The importance of human factors/usability in the design of engineering systems goes back to 1901; the Army Signal Corps contract document for the development of the Wright Brothers' airplane quite clearly stated that the aircraft be "Simple to operate and maintain."[6] Nonetheless, human factors as a technical discipline emerged only after World War II, basically due to the military systems' increasing complexity, as well as the critical human role in operating them.

In 1957, the Human Factors Society of America was incorporated, and in 1983, the Association for Computing Machinery (AMC) Special Interest Group on Computer and Human Interaction (SIGCHI) was formed [7,8]. In the mid-1980s, the term *usability engineering* was coined [9,10]. Additional information on the history of usability engineering is available in Reference 8.

1.2 Systems reliability and usability-related facts, figures, and examples

Some of the facts, figures, and examples concerned with systems reliability and usability, either directly or indirectly, are as follows:

- Decade breakdowns of worldwide fatal commercial aircraft accidents due to mechanical failure during the period 1950–2008 are 21 (1950–1959), 20 (1960–1969), 23 (1970–1979), 21(1980–1989), 21 (1990–1999), and 28 (2000–2008) [11].

- As per Reference 12, it costs about $100 billion per year in lost productivity to American businesses because office workers "futz" with their machines an average of 5.1 hours per week.
- Some studies carried out in Japan clearly indicate that more than 50% of working accidents with robots can be attributed to faults in the control system's electronic circuits [13].
- A study reported that operator-related error accounts for over 50% of all technical medical equipment–related problems [14].
- In 2002, a study commissioned by the National Institute of Standards and Technology (NIST) reported that software errors cost the United States economy around $59 billion annually [15].
- As per Reference 16, the Center for Devices and Radiological Health (CDRH) of the U.S. Food and Drug Administration (FDA) reported that around 60% of the deaths and serious injuries associated with medical devices occurred due to user error.
- It is estimated that approximately 80% of the cost of software maintenance is related to unforeseen/unmet user needs [24].
- As per Reference 17, it is estimated that around 80% of software maintenance–related cost is due to unforeseen/unmet user needs.
- A worker holding the robot controls in his or her hand activated the robot while bending over the wheel to check the settings. The robot pinned the worker against the wheel and crushed him or her to death [18–23].
- As per Reference 24, in 1996, the direct cost of corrosion-related failures, including maintenance in the U.S. petroleum industrial sector, was about $3.7 billion per year.
- A study reported that approximately 63% of all software-related projects overran their predicted costs, with the top four principal reasons all related to usability [25].
- As per References 26,27, the Internet has grown from 4 hosts in 1969 to over 147 million hosts and 38 sites in 2002, and in 2001, there were over 52,000 Internet-related incidents and failures.
- A study commissioned by the U.S. Nuclear Regulatory Commission reported that around 65% of nuclear system failures involve human error to a degree [28].
- As per Reference 29, with the application of usability engineering principles, 33%–50% reduction in the product development cycle occurred.
- A study reported that a usability engineering product achieved 80% greater revenue than the first release, which was developed without consideration of usability engineering principles [30,31].

- As per Reference 32, in 1991, United Airlines Flight 585 (aircraft type: Boeing 737–291) crashed due to rudder device malfunction, causing 25 fatalities.
- A study reported that because computers are often quite difficult to use, organizations and companies frequently provide approximately $3,150 worth of technical support for each user of the equipment [33].
- As per Reference 34, a study reported that usability-associated design changes at IBM resulted in the reduction of 9.6 minutes per task, which translates into a projected internal savings of $6.8 million for the entire organization for the year 1991 alone.

1.3 Terms and definitions

There are a large number of terms and definitions used in the area of reliability and usability. Some of these are as follows [8,35–40]:

- *Reliability*: The probability that an item will perform its stated mission satisfactorily for the specified time period when used according to specified conditions
- *Usability*: The quality of an interactive system with regard to factors such as ease of use, ease of learning, and ease of satisfaction
- *Failure*: The inability of an item to function within specified guidelines
- *Mission time*: The element of uptime that is needed to carry out a specified mission profile
- *Redundancy*: The existence of more than one means for carrying out a specified function
- *Availability*: The probability that an item or system is available for application or use when needed
- *Continuous task*: A task that involves some kind of tracking activity (e.g., monitoring a changing condition or situation)
- *Downtime*: The time period during which the item or system is not in the necessary condition to perform its specified mission
- *Human error*: The failure to perform a specified task (or the performance of a forbidden action) that could result in disruption of scheduled operations or damage to equipment and property
- *Failure mode*: The abnormality of item or system performance that causes the item or system to be considered as failed
- *Random failure*: Any failure whose occurrence cannot be predicted
- *User interface*: The physical representations and procedures for viewing and interacting with the product or system functionality
- *User task*: A desired result of activities that the product or system user would like to accomplish

- *Usability engineering*: Iterative design and evaluation for providing customer feedback on the usefulness and usability of a product's or system's design and functionality throughout the development phase
- *Usability evaluation*: Any analytical or empirical activity directed at assessing or understanding the usability of an interactive product/ system
- *Usability inspection*: An analytical approach in which usability specialists and experts evaluate the user interaction needed for carrying out pivotal or crucial tasks with an interactive system or product for determining the problematic aspects of user input or system response

1.4 Useful sources for obtaining information on reliability and usability

There are many sources for obtaining information, directly or indirectly, concerned with systems reliability and usability. Some of the sources considered most useful are presented in the following sections, classified in a number of distinct categories.

1.4.1 Journals and magazines

- *IEEE Transactions on Reliability*
- *International Journal of Reliability, Quality, and Safety Engineering*
- *Reliability Engineering and System Safety*
- *Microelectronics and Reliability*
- *Engineering Failure Analysis*
- *Ergonomics*
- *ACM Interactions*
- *Human-Computer Interaction*
- *User Modeling and User-Adapted Interaction (UMUAI)*
- *Interacting with Computers*
- *International Journal of Human-Computer Studies*
- *Journal of Multimodel Use Interfaces*
- *ACM Transactions on Computer-Human Interaction (TOCHI)*
- *Journal of Usability Studies*

1.4.2 Conference proceedings

- *Proceedings of the Annual Reliability and Maintainability Symposium*
- *Proceedings of the European Conferences on Safety and Reliability*
- *Proceedings of the ISSAT International Conferences on Reliability and Quality in Design*

- *Proceedings of the Human Factors and Ergonomics Society Annual Meetings*
- *Proceedings of the Conferences on Advances in Usability Engineering*
- *Proceedings of the Conferences on Human Factors in Computing Systems*

1.4.3   Books

- Shooman, M. L., *Probabilistic Reliability: An Engineering Approach,* McGraw-Hill Book Company, New York, 1968.
- Cox, S. J., *Reliability, Safety, and Risk Management: An Integrated Approach,* Butterworth-Heinemann, New York, 1991.
- Dhillon, B. S., *Design Reliability: Fundamentals and Applications,* CRC Press, Boca Raton, FL, 1999.
- Evans, J. W., and Evans, J. Y., *Product Integrity and Reliability in Design,* Springer-Verlag, New York, 2001.
- Dhillon, B. S., *Computer System Reliability: Safety and Usability,* CRC Press, Boca Raton, FL, 2013.
- Bawa, J., Dorazio, P., and Trenner, L., editors, *The Usability Business: Making the Web Work,* Springer-Verlag, New York, 2001.
- Mayhew, D. J., *The Usability Engineering Lifecycle: A Practitioner's Handbook for User Interface Design,* Morgan Kaufmann Publishers, San Francisco, 1999.
- Nielsen, J., *Usability Engineering,* Academic Press, Boston, 1993.
- Rosson, M. B., and Carroll, J. M., *Usability Engineering: Scenario-Based Development of Human-Computer Interaction,* Academic Press, San Francisco, 2002.
- Rubin, J., *Handbook of Usability Testing: How to Plan, Design, and Conduct Effective Tests,* John Wiley and Sons, New York, 1994.
- Adler, P. S., and Winograd, T. A., editors, *Usability: Turning Technologies into Tools,* Oxford University Press, New York, 1992.
- Wiklund, M. E., editor, *Usability in Practice: How Companies Develop User-Friendly Products,* AP Professional, Inc., Boston, 1994.
- Dhillon, B. S., *Engineering Usability: Fundamentals, Applications, Human Factors, and Human Error,* American Scientific Publishers, Stevenson Ranch, CA, 2004.

1.4.4   Standards

- MIL-STD-721, Definitions of Terms for Reliability and Maintainability, U.S. Department of Defense, Washington, DC.
- MIL-STD-1908, Definitions of Human Factors Terms, U.S. Department of Defense, Washington, DC.
- MIL-HDBK-217, Reliability Prediction of Electronic Equipment, U.S. Department of Defense, Washington, DC.

- MIL-STD-785, Reliability Program for Systems and Equipment, Development and Production, U.S. Department of Defense, Washington, DC.
- MIL-STD-756, Reliability Modeling and Prediction, U.S. Department of Defense, Washington, DC.
- MIL-STD-1629, Procedures for Performing Failure Mode, Effects and Criticality Analysis, U.S. Department of Defense, Washington, DC.
- MIL-HDBK-338, Electronics Reliability Design Handbook, U.S. Department of Defense, Washington, DC.
- MIL-STD-790, Reliability Assurance Program for Electronic Parts Specifications, U.S. Department of Defense, Washington, DC.
- MIL-STD-2155, Failure Reporting, Analysis, and Corrective Action (FRACAS), U.S. Department of Defense, Washington, DC.

1.4.5   Data sources

- Government Industry Data Exchange Program (GIDEP), GIDEP Operations Center, U.S. Department of Navy, Corona, CA.
- National Technical Information Service (NTIS), United States Department of Commerce, Springfield, VA.
- Reliability Analysis Center, Rome Air Development Center (RADC), Griffiss Air Force Base, Rome, NY.
- Defense Technical Information Center, DTIC-FDAC, Fort Belvoir, VA.
- Computer Accident/Incident Report System, System Safety Development Center, Idaho Falls, OH.
- American National Standards Institute (ANSI), New York.

1.4.6   Organizations

- Reliability Society, Institute of Electrical and Electronics Engineers (IEEE), Piscataway, NJ.
- Usability Professionals' Association, Bloomingdale, IL.
- Human Factors and Ergonomics Society, Santa Monica, CA.
- Association for Computing Machinery (ACM) Special Group on Computer and Human Interaction (SIGCHI), New York.
- Society for Maintenance and Reliability Professionals, Chicago.

1.5   Scope of this book

Nowadays, engineering systems are an important element of the world economy, and each year, billions of dollars are spent to develop, manufacture, operate, and maintain various types of engineering systems. The reliability and usability of these systems have become more important than ever because of their increasing complexity, sophistication, and nonspecialist users. Over the years, a large number of journal and conference proceedings articles, technical reports, and other publications on the reliability and usability of engineering systems have appeared in the literature. However, to the best of the author's knowledge, there is no book that covers the topics of reliability and usability within its framework. This is a significant impediment to information seekers on these topics because they have to consult various sources.

Thus, the main objectives of this book are (1) to eliminate the need for professionals and others concerned with engineering system reliability and usability to consult diverse sources in obtaining the desired information and (2) to provide up-to-date information on the subject. This book will be useful to many individuals, including design engineers, system engineers, reliability specialists, usability specialists, human factors and ergonomics specialists, computer system engineers, computer-interface specialists, engineering undergraduate and graduate students, researchers, and instructors in the area of reliability and usability engineering, and engineers at large.

PROBLEMS

1. Write an essay on the reliability and usability of engineering systems.
2. Define the following four terms:
    i. Reliability
    ii. Usability
    iii. Human error
    iv. Usability evaluation
3. List seven important facts and figures concerning engineering system reliability and usability.
4. List six of the most important journals or magazines for obtaining information on reliability or usability.
5. List at least four data information sources.
6. List at least four organizations that are concerned with reliability/usability.
7. List four important reliability standards.
8. Define the following terms:
    i. Mission time
    ii. Continuous task
    iii. Usability engineering
9. What is the difference between usability evaluation and usability inspection?
10. Define the following terms:
    i. User task
    ii. Downtime
    iii. User interface

References

1. Smith, S. A., Service Reliability Measured by Probabilities of Outage, *Electrical World*, Vol. 103, 1934, pp. 371–374.
2. Lyman, W. J., Fundamental Consideration in Preparing a Master System Plan, *Electrical World*, Vol. 101, 1933, pp. 778–792.
3. Coppola, A., Reliability Engineering of Electronic Equipment: A Historical Perspective, *IEEE Transactions on Reliability*, Vol. 33, 1984, pp. 29–35.
4. Dhillon, B. S., *Design Reliability: Fundamentals and Applications*, CRC Press, Boca Raton, FL, 1999.
5. Dhillon, B. S., *Reliability and Quality Control: Bibliography on General and Specialized Areas*, Beta Publishers, Gloucester, Ontario, Canada, 1992.
6. AMCP 706-133. *Engineering Design Handbook: Maintainability Engineering Theory and Practice*, Department of Defense, Washington, DC, 1976.
7. Shackel, B., and Richardson, S., Human Factors for Informatics Usability: Background and Overview, in *Human Factors for Informatics Usability*, edited by B. Shackel and S. Richardson, Cambridge University Press, Cambridge, UK, 1991, pp. 1–19.
8. Dhillon, B. S., *Engineering Usability: Fundamentals, Applications, Human Factors, and Human Error*, American Scientific Publishers, Stevenson Ranch, CA, 2004.
9. Rosson, M. B., and Carroll, J. M., *Usability Engineering: Scenario-Based Development of Human-Computer Interaction*, Academic Press, San Francisco, 2002.
10. Butler, K. A., Usability Engineering Turns Ten, *Interactions*, January 1996, pp. 59–75.
11. Dhillon, B. S., *Transportation Systems Reliability and Safety*, CRC Press, Boca Raton, FL, 2011.
12. SBT Accounting Systems, Westlake Consulting Company, Inc., Houston, 1997.
13. Retsch, T., Schmitter, G., and Marty, A., Safety Principles for Industrial Robots, in *Encyclopaedia of Occupational Health and Safety*, Vol. II, edited by J. M. Stellman, International Labor Organization, Geneva, Switzerland, 2011, pp. 58.56–58.58.
14. Dhillon, B. S., Reliability Technology in Healthcare Systems, Proceedings of the IASTED International Symposium on Computers and Advanced Technology in Medicine, Health Care, and Bioengineering, 1990, pp. 84–87.
15. National Institute of Standards and Technology (NIST), Gaithersburg, MD, 2002.
16. Bogner, M. S., Medical Devices: A New Frontier for Human Factors, *CSERIAC Gateway*, Vol. IV, No. 1, 1993, pp. 12–14.
17. Pressman, R. S., *Software Engineering: A Practioner's Approach*, McGraw-Hill, New York, 1992.
18. Dhillon, B. S., *Robot Reliability and Safety*, Springer-Verlag, New York, 1991.
19. *Study on Accidents Involving Industrial Robots*, Report No. PB 83239822, Prepared by the Japanese Ministry of Labor, Tokyo, 1982. Available from the National Technical Information Service (NTIS), Springfield, VA.
20. Altamuro, V. M., Working Safely with the Iron Collar Worker, *National Safety News*, July 1983, pp. 38–40.

21. Nicolaisen, P., Safety Problems Related to Robots, *Robotics*, Vol. 3, 1987, pp. 205–211.

22. Report No. 0552652, Occupational Safety and Health Administration (OSHA), Washington, DC., October 10, 2006.

23. Lauch, K. E., New Standards for Industrial Robot Safety, *CIM Review*, Spring, 1986, pp. 60–68.

24. Kane, R. D., Corrosion in Petroleum Refining and Petrochemical Operations, in *Metals Handbook, Vol. 13C: Environments and Industries*, edited by S. O. Cramer and B. S. Covino, ASM International, Metals Park, OH, 2003, pp. 967–1014.

25. Landauer, A. L., Prasad, J., Nine Management Guidelines for Better Cost Estimating, *Communications of the ACM*, Vol. 35, No. 2, 1992, pp. 51–59.

26. Hafner, K., and Lyon, M., *Where Wizards Stay up Late: The Origin of the Internet*, Simon and Schuster, New York, 1996.

27. Dhillon, B. S., *Applied Reliability and Quality: Fundamentals, Methods, and Procedures*, Springer, London, 2007.

28. Trager, T. A., *Case Study Report on Loss of Safety System Function Events, Report No. AEOD/C 504*, United States Nuclear Regulatory Commission (NRC), Washington, DC, 1985.

29. Bosert, J. L., *Quality Function Deployment: A Practitioner's Approach*, American Society for Quality Control (ASQC) Quality Press, New York, 1991.

30. Wixon, D., Jones, S., Usability for Fun and Profit: A Case Study of the Redesign of the VAX RALLY, in *Human-Computer Interface Design: Success Stories, Emerging Methods, and Real-World Context*, edited by M. Rudisill, C. Lewis, P. G. Polson, and T. McKay, Morgan Kaufmann Publishers, San Francisco, 1995.

31. Bevan, N., *Cost Benefit Analysis*, Report No. 3 version 1.1, September 8, 2000. Serco Usability Services, Alderney House, 4 Sandy Lane, Teddington, Middx, UK.

32. *Aircraft Accident Report: United Airlines Flight 585*, Report No. AAR92-06, National Transportation Safety Board, Washington, DC, 1992.

33. Gibbs, W. W., Taking Computer to Task, *Scientific American*, No. 7, 1997, pp. 10–11.

34. Karat, C., Cost-Benefit Analysis of Usability Engineering Techniques, Proceedings of the Human Factors Society Conference, 1990, pp. 839–843.

35. MIL-STD-721, *Definitions of Effectiveness Terms for Reliability, Maintainability, Human Factors, and Safety*, Department of Defense, Washington, DC.

36. Omdahl, T. P., editor, *Reliability, Availability, and Maintainability (RAM) Dictionary*, ASQC Quality Press, Milwaukee, 1988.

37. Neresky, J. J., Reliability Definitions, *IEEE Transactions on Reliability*, Vol. 19, 1970, pp. 198–200.

38. McKenna, T., and Oliverson, R., *Glossary of Reliability and Maintenance Terms*, Gulf Publishing Company, Houston, 1997.

39. Glossary of Terms used in Usability Engineering, Available online at http://www.ucc.ie/hfrg/baseline/glossary.html.

40. *User-Centered Design Process for Interactive Systems*, ISO 13407 1999, International Organization for Standardization (ISO), Geneva, Switzerland, 1999.

chapter two

Basic mathematical concepts

2.1 Introduction

Just as in the development of other areas of science and technology, mathematics has played an important role in the development of reliability and usability fields. The origin of the word *mathematics* may be traced to the ancient Greek word *mathema*, which basically means "science, learning, or knowledge," and the history of our current number symbols, frequently referred to as a Hindu-Arabic numeral system, goes back to around 250 B.C., to the stone columns erected by Asoka, a Scythian emperor of India [1]. More clearly, the evidence of the use of these number symbols can be seen in notches found on the stone columns.

The earliest reference to the probability concept may be traced to a gambler's manual written by Girolamo Cardano (1501–1576) [2]. However, Pierre Fermat (1601–1665) and Blaise Pascal (1623–1662) were the first two individuals who solved, correctly and independently, the problem of dividing the winnings in a game of chance [1,2]. Boolean algebra, which plays a very important role in modern probability theory, is named after George Boole (1815–1864), an English mathematician who published in 1847 a pamphlet titled "The Mathematical Analysis of Logic: Being an Essay Towards a Calculus of Deductive Reasoning" [1–2].

Laplace transforms, often used in the area of reliability for finding solutions to first-order linear differential equations, were developed by a French mathematician named Pierre-Simon Laplace (1749–1827). Additional information on the history of mathematics and probability is available in References 1,2. This chapter presents the basic mathematical concepts that are considered useful to understand the subsequent chapters of this book.

2.2 Arithmetic mean and mean deviation

A set of given system reliability or usability data is useful only if it is analyzed properly. More specifically, the data have certain characteristics that are useful to describe the nature of a given data set, thus allowing one to make better decisions. This section presents two statistical measures that are useful for studying system reliability- and usability-related data [3–5].

2.2.1   *Arithmetic mean*

Often, the *arithmetic mean* is referred to simply as the *mean* and is expressed by

$$m = \frac{\sum_{i=1}^{k} x_i}{k}, \tag{2.1}$$

where m is the mean value (i.e., arithmetic mean); k is the number of data values; and x_i is the data value i, for $i = 1, 2, 3, \ldots, k$.

EXAMPLE 2.1

Assume that the inspection department of an organization involved in the manufacture of various types of engineering systems inspected seven identical systems and discovered 2, 5, 3, 4, 1, 4, and 2 defects in each one. Calculate the average number of defects (i.e., the arithmetic mean) per engineering system.

By substituting the specified data values into Eq. (2.1), we get

$$m = \frac{2 + 5 + 3 + 4 + 1 + 4 + 2}{7} = 3.$$

Thus, the average number of defects per engineering system is 3. In other words, the arithmetic mean of the given data set is 3.

2.2.2   *Mean deviation*

A *mean deviation* is a measure of dispersion whose value indicates the degree to which a given data set tends to spread about a mean value. Mean deviation is expressed by

$$MD = \frac{\sum_{i=1}^{n} |x_i - m|}{n}, \tag{2.2}$$

where MD is the mean deviation; n is the number of data values; m is the mean value of the given data set; x_i is the data value i, for $i = 1, 2, 3, \ldots, n$; and $|x_i - m|$ is the absolute value of the deviation of x_i from m.

EXAMPLE 2.2

Calculate the mean deviation of the data set provided in Example 2.1.

In Example 2.1, the calculated mean value (i.e., arithmetic mean) of the given data set is 3 defects per engineering system. Thus, using this calculated value and the given data values in Equation 2.2, we get

$$MD = \frac{|2-3| + |5-3| + 3-3| + |4-3| + |1-3| + |4-3| + |2-3|}{7}$$

$$= \frac{[1+2+0+1+2+1+1]}{7}$$

$$= 1.1428.$$

Thus, the mean deviation of the data set in Example 2.1 is 1.1428.

2.3   Boolean algebra laws

Boolean algebra, named after its founder, George Boole, is used to some extent in studies concerning system reliability and usability. Some of its laws that are considered useful for understanding subsequent chapters of this book are presented as follows [3,4,6–8]:

$$A.B = B.A, \tag{2.3}$$

where A is an arbitrary set or event; B is an arbitrary set or event; and Dot(.) denotes the intersection of sets.

It is to be noted that when Eq. (2.3) is written without the dot (e.g., AB), it still conveys the same meaning.

$$A + B = B + A, \tag{2.4}$$

where + indicates the union of sets or events.

$$A + A = A \tag{2.5}$$

$$AA = A \tag{2.6}$$

$$A + AB = A \tag{2.7}$$

$$B(B + A) = B \tag{2.8}$$

$$A(B + C) = AB + AC, \tag{2.9}$$

where C is an arbitrary set or event.

$$(A + B)(A + C) = A + BC \tag{2.10}$$

$$(A + B) + C = A + (B + C) \tag{2.11}$$

$$(AB)C = A(BC) \tag{2.12}$$

It is to be noted that in the published literature, Equations 2.3 and 2.4 are known as commutative law, Equations 2.5 and 2.6 as idempotent law, Equations 2.7 and 2.8 as absorption law, Equations 2.9 and 2.10 as distributive law, and Equations 2.11 and 2,12 as associative law [9].

2.4   Probability definition and properties

Probability is defined as follows [10]:

$$P(X) = \lim_{n \to \infty} \left(\frac{N}{n} \right), \tag{2.13}$$

where $P(X)$ is the probability of occurrence of event X and N is the number of times that event X occurs in experiments repeated n times.

Some of the basic probability properties are as follows [7–10]:

- The probability of the occurrence of an event (say Y) is

$$0 \leq P(Y) \leq 1. \tag{2.14}$$

- The probability of the occurrence and nonoccurrence of an event (Y) is always

$$P(Y) + P(\bar{Y}) = 1, \tag{2.15}$$

where $P(Y)$ is the probability of the occurrence of event Y and $P(\bar{Y})$ is the probability of the nonoccurrence of event Y.

- The probability of an intersection of n independent events is

$$P(Y_1 Y_2 Y_3 \ldots Y_n) = P(Y_1)P(Y_2)P(Y_3) \ldots P(Y_n), \tag{2.16}$$

where $P(Y_i)$ is the occurrence probability of event Y_i for $i = 1, 2, 3, \ldots, n$.

- The probability of the union of n independent events is

$$P(Y_1 + Y_2 + \cdots + Y_n) = 1 - \prod_{i=1}^{n} (1 - P(Y_i)) \tag{2.17}$$

- The probability of the union of n mutually exclusive events is

$$P(Y_1 + Y_2 + \cdots + Y_n) = \sum_{i=1}^{n} P(Y_i). \tag{2.18}$$

EXAMPLE 2.3

Assume that an engineering system has two critical subsystems, Y_1 and Y_2. The failure of either subsystem can cause an accident, directly or indirectly. The failure probability of subsystems Y_1 and Y_2 is 0.04 and 0.06, respectively.

Calculate the probability of the occurrence of an accident in the engineering system if both subsystems fail independently.

By substituting the given data values into Equation 2.17, we get

$$P(Y_1 + Y_2) = 1 - \prod_{i=1}^{2} (1 - P(Y_i))$$

$$= P(Y_1) + P(Y_2) - P(Y_1)P(Y_2)$$
$$= 0.04 + 0.06 - (0.04)(0.06)$$
$$= 0.0976.$$

Thus, the probability of the occurrence of an accident in the engineering system is 0.0976.

2.5 Useful definitions

This section presents a number of mathematical definitions that are considered very useful for performing various types of system reliability and usability studies.

2.5.1 Cumulative distribution function

For continuous random variables, the cumulative distribution function is defined by Reference 10 as follows:

$$F(t) = \int_{-\infty}^{t} f(x)dx, \tag{2.19}$$

where x is a continuous random variable; $f(x)$ is the probability density function; and $F(t)$ is the cumulative distribution function.

For $t = \infty$, Eq. (2.19) yields

$$F(\infty) = \int_{-\infty}^{\infty} f(x)dx \tag{2.20}$$

$$= 1.$$

This means that the total area under the probability density curve is equal to unity.

It is to be noted that usually in reliability and usability studies of engineering systems, Equation 2.19 is written simply, as follows:

$$F(t) = \int_0^t f(x)dx. \tag{2.21}$$

2.5.2 Probability density function

For a continuous random variable, this is defined by Reference 10 as

$$f(t) = \frac{dF(t)}{dt}, \tag{2.22}$$

where $f(t)$ is the probability density function and $F(t)$ is the cumulative distribution function.

2.5.3 Expected value

The expected value of a continuous random variable is expressed by

$$E(t) = m = \int_{-\infty}^{\infty} tf(t)dt \tag{2.23}$$

where $E(t)$ is the expected value of the continuous random variable t and m is the mean value.

2.5.4 Variance

The variance of a random variable t is expressed by

$$\sigma(t) = E(t^2) - [E(t)]^2 \tag{2.24}$$

or

$$\sigma(t) = \int_0^{\infty} t^2 f(t)dt - m^2, \tag{2.25}$$

where $\sigma(t)$ is the variance of a random variable t and m is the mean value.

2.5.5 Laplace transform

The Laplace transform of the function $f(t)$ is expressed by

$$f(s) = \int_0^\infty f(t)e^{-st}dt, \tag{2.26}$$

where $f(s)$ is the Laplace transform of $f(t)$, t is the time variable, and s is the Laplace transform variable.

EXAMPLE 2.4

Obtain the Laplace transform of the following function:

$$f(t) = e^{-\theta t}, \tag{2.27}$$

where θ is a constant.

By inserting Equation 2.27 into Equation 2.26, we obtain

$$f(s) = \int_0^\infty e^{-\theta t}e^{-st}dt$$

$$= \int_0^\infty e^{-(s+\theta)t}dt \tag{2.28}$$

$$= \frac{1}{s+\theta}.$$

Laplace transforms of several functions that occur frequently in the analysis of system reliability and usability are presented in Table 2.1 [11,12].

Table 2.1 Laplace transforms of several functions

No.	$f(t)$	$f(s)$
1	k (a constant)	$\dfrac{k}{s}$
2	$e^{-\lambda t}$	$\dfrac{1}{s+\lambda}$
3	$\dfrac{df(t)}{dt}$	$sf(s) - f(0)$
4	t^n, for $n = 0,1,2,3,\ldots$	$\dfrac{n!}{s^{n+1}}$
5	$\theta_1 f_1(t) + \theta_2 f_2(t)$	$\theta_1 f_1(s) + \theta_2 f_2(s)$
6	$te^{-\lambda t}$	$\dfrac{1}{(s+\lambda)^2}$
7	$tf(t)$	$-\dfrac{df(s)}{ds}$

2.5.6   *Laplace transform: Final-value theorem*

If the following limits exist, then the final-value theorem may be defined as

$$\lim_{t \to \infty} f(t) = \lim_{s \to 0} [sf(s)]. \tag{2.29}$$

EXAMPLE 2.5

Prove, by using the following equation, that the left side of Equation 2.29 is equal to its right side:

$$f(t) = \frac{\alpha_1}{(\alpha_1 + \alpha_2)} + \frac{\alpha_2}{(\alpha_1 + \alpha_2)} e^{-(\alpha_1 + \alpha_2)t}, \tag{2.30}$$

where α_1 and α_2 are constants.

By inserting Equation 2.30 into the left side of Equation 2.29, we obtain

$$\lim_{t \to \infty} \left[\frac{\alpha_1}{(\alpha_1 + \alpha_2)} + \frac{\alpha_2}{(\alpha_1 + \alpha_2)} e^{-(\alpha_1 + \alpha_2)t} \right] = \frac{\alpha_1}{(\alpha_1 + \alpha_2)}. \tag{2.31}$$

Using the elements in Table 2.1, we obtain the following Laplace transform of Equation 2.30:

$$f(s) = \frac{\alpha_1}{s(\alpha_1 + \alpha_2)} + \frac{\alpha_2}{(\alpha_1 + \alpha_2)} \cdot \frac{1}{(s + \alpha_1 + \alpha_2)}. \tag{2.32}$$

By substituting Equation 2.32 into the right side of Equation 2.29, we obtain

$$\lim_{s \to 0} s \left[\frac{\alpha_1}{s(\alpha_1 + \alpha_2)} + \frac{\alpha_2}{(\alpha_1 + \alpha_2)} \cdot \frac{1}{(s + \alpha_1 + \alpha_2)} \right] = \frac{\alpha_1}{(\alpha_1 + \alpha_2)}. \tag{2.33}$$

The right sides of Equations 2.31 and 2.33 are the same. This result proves that the left side of Equation 2.29 is equal to its right side.

2.6   *Probability distributions*

Although there are many probability or statistical distributions in the published research literature, this section presents five of these distributions, which are considered very useful for performing various types of system reliability and usability studies [13–15].

2.6.1 Exponential distribution

The exponential distribution is one of the simplest continuous random variable distributions frequently used in the industry, particularly for conducting reliability studies [16]. The probability density function of the distribution is defined by

$$f(t) = \alpha e^{-\alpha t}, \text{ for } \alpha > 0, t \geq 0,$$ (2.34)

where $f(t)$ is the probability density function; t is the time (i.e., a continuous random variable); and α is the distribution parameter.

By inserting Equation 2.34 into Equation 2.19, we obtain the following expression for the cumulative distribution function:

$$F(t) = 1 - e^{-\alpha t}.$$ (2.35)

Using Equations 2.23 and 2.34, we obtain the following equation for the distribution expected value (i.e., mean value):

$$E(t) = m = \frac{1}{\alpha}.$$ (2.36)

EXAMPLE 2.6

Assume that the mean time to failure of an engineering system is 800 h. Calculate the engineering system's probability of failure during a 500-h mission by using Equations 2.35 and 2.36.

By inserting the specified data value into Equation 2.36, we get

$$\alpha = \frac{1}{m} = \frac{1}{800} = 0.00125 \text{ failures per hour}$$

By substituting the calculated and the specified data values into Equation 2.35, we get

$$F(500) = 1 - e^{-(0.00125)(500)}$$
$$= 0.4647.$$

Thus, the engineering system's probability of failure during the 500-h mission is 0.4647.

2.6.2 Binomial distribution

The binomial distribution is a discrete random variable distribution, and it is used in circumstances where one is concerned with the probabilities of a particular outcome, such as the number of occurrences (e.g., failures) in a sequence of k trials. More clearly, each trial has two possible outcomes (e.g., success or failure), but the probability of each trial remains unchanged or constant. It is to be noted that this distribution is also known as the *Bernoulli distribution*, after Jakob Bernoulli (1654–1705) [1]. The binomial probability density function, $f(y)$, is expressed by

$$f(y) = \binom{k}{j} p^y q^{k-y}, \text{ for } y = 0, 1, 2, \ldots, k, \tag{2.37}$$

where $\binom{k}{j} = \frac{k!}{j!(k-j)!}$, y is the number of nonoccurrences (e.g., failures) in k trials, p is the single trial probability of occurrence (e.g., success), and q is the single trial probability of nonoccurrence (e.g., failure).

The cumulative distribution function is given by

$$F(y) = \sum_{j=0}^{y} \binom{k}{j} p^j q^{k-j}. \tag{2.38}$$

where $F(y)$ is the cumulative distribution function or the probability of y or fewer non occurrences (e.g., failures) in k trials.

2.6.3 Rayleigh distribution

The Rayleigh distribution is a continuous random variable probability distribution named after its founder, John Raleigh (1842–1919), [1] and its probability density function is defined by

$$f(t) = \left(\frac{1}{\theta^2}\right) t e^{-(t/\theta)^2}, \text{ for } \theta \rangle 0, \ t \geq 0, \tag{2.39}$$

where θ is the distribution parameter.

By substituting Equation 2.39 into Equation 2.19, we obtain the following equation for the cumulative distribution function:

$$F(t) = 1 - e^{-(t/\theta)^2}. \tag{2.40}$$

By inserting Equation 2.39 into Equation 2.23, we get the following equation for the distribution expected value (i.e., mean value):

$$E(t) = \theta \ \Gamma\left(\frac{3}{2}\right), \tag{2.41}$$

where $\Gamma(.)$ is the gamma function, which is defined by

$$\Gamma(k) = \int_0^\infty t^{k-1}e^{-t}dt, \text{ for } k > 0. \tag{2.42}$$

2.6.4   Weibull distribution

The Weibull distribution is a continuous random variable distribution named after Walliodi Weibull, a Swedish mechanical engineering professor, who developed it in the early 1950s [17]. The probability density function of the distribution is defined by

$$f(t) = \frac{bt^{b-1}}{\theta^b}e^{-(t/\theta)^b}, \text{ for } \theta > 0, \ b > 0, \ t \geq 0, \tag{2.43}$$

where b and θ are the distribution shape and scale parameters, respectively.

By inserting Equation 2.43 into Equation 2.19, we obtain the following equation for the cumulative distribution function:

$$F(t) = 1 - e^{-(t/\theta)^b}. \tag{2.44}$$

By substituting Equation 2.43 into Equation 2.23, we obtain the following equation for the distribution expected value (i.e., mean value):

$$E(t) = \theta \ \ \Gamma\left(1 + \frac{1}{b}\right). \tag{2.45}$$

Finally, note that the exponential and Rayleigh distributions are special cases of the Weibull distribution for $b = 1$ and $b = 2$, respectively.

2.6.5   Bathtub hazard rate curve distribution

The bathtub hazard rate curve distribution is a continuous random variable distribution developed in 1981 [18]. In the published literature by other authors around the world, it is generally referred to as the *Dhillon distribution/law/model* [19–43]. The distribution can assume a bathtub shape, increasing and decreasing the hazard rates.

The probability density function of the distribution is defined by Reference 18 as follows:

$$f(t) = b\theta(\theta t)^{b-1}e^{-\left\{e^{(\theta t)^b}-(\theta t)^b-1\right\}}, \text{ for } t \geq 0, \ \theta > 0, \ b > 0, \tag{2.46}$$

where θ and b are the distribution scale and shape parameters, respectively.

By substituting Equation 2.46 into Equation 2.19, we get the following equation for the cumulative distribution function:

$$F(t) = 1 - e^{-\{e^{(\theta t)^b} - 1\}}.\qquad(2.47)$$

Finally, note that for $b = 0.5$, this probability distribution gives the bathtub-shaped hazard rate curve, and for $b = 1$, it gives an extreme value probability distribution. More clearly, the extreme value probability distribution is a special case of this probability distribution at $b = 1$.

2.7 Solving first-order differential equations using Laplace transforms

Usually, Laplace transforms are used to find solutions to first-order linear differential equations in reliability and usability analysis-related studies of various types of systems. The example presented here demonstrates the finding of solutions to a set of linear first-order differential equations, describing a system with respect to reliability and usability using Laplace transforms.

EXAMPLE 2.7

Assume that an engineering system can be in any of these three states: operating normally, failed due to hardware failure, or failed due to user error. The following three first-order linear differential equations describe the engineering system under consideration:

$$\frac{dP_0(t)}{dt} + (\lambda_h + \lambda_u)P_0(t) = 0\qquad(2.48)$$

$$\frac{dP_1(t)}{dt} - \lambda_h P_0(t) = 0\qquad(2.49)$$

$$\frac{dP_2(t)}{dt} - \lambda_u P_0(t) = 0,\qquad(2.50)$$

where λ_h is the engineering system constant hardware failure rate, λ_u is the engineering system constant user error rate, and $P_i(t)$ is the probability that the engineering system is in state i at time t, for $i = 0$ (operating normally), $i = 1$ (failed due to hardware failure), and $i = 2$ (failed due to user error).

At time $t = 0$, $P_0(0) = 1$, $P_1(0) = 0$, and $P_2(0) = 0$.

Solve the differential Equations 2.48 through 2.50 by using Laplace transforms.

Using the information in Table 2.1, Equations 2.48 through 2.50, and the given initial conditions, we obtain

$$sP_0(s) - 1 + (\lambda_h + \lambda_u)P_0(s) = 0 \tag{2.51}$$

$$sP_1(s) - \lambda_h P_0(s) = 0 \tag{2.52}$$

$$sP_2(s) - \lambda_u P_0(s) = 0. \tag{2.53}$$

By solving Equations 2.51 through 2.53, we get

$$P_0(s) = \frac{1}{(s + \lambda_h + \lambda_u)} \tag{2.54}$$

$$P_1(s) = \frac{\lambda_h}{s(s + \lambda_h + \lambda_u)} \tag{2.55}$$

$$P_2(s) = \frac{\lambda_u}{s(s + \lambda_h + \lambda_u)}. \tag{2.56}$$

By taking the inverse Laplace transforms of Equations 2.54 through 2.56, we obtain

$$P_0(t) = e^{-(\lambda_h + \lambda_u)t} \tag{2.57}$$

$$P_1(t) = \frac{\lambda_h}{(\lambda_h + \lambda_u)}[1 - e^{-(\lambda_h + \lambda_u)t}] \tag{2.58}$$

$$P_2(t) = \frac{\lambda_u}{(\lambda_h + \lambda_u)}[1 - e^{-(\lambda_h + \lambda_u)t}]. \tag{2.59}$$

Thus, Equations 2.57 through 2.59 are the solutions to Equations 2.48 through 2.50.

PROBLEMS

1. Assume that the quality control department of an organization involved in the manufacture of various types of engineering systems inspected 10 identical systems and found 3, 1, 5, 2, 4, 7, 2, 1, 2, and 3 defects in each system. Calculate the average number of defects (i.e., the arithmetic mean) per engineering system.
2. Calculate the mean deviation of the data set given in Question 1.
3. What is idempotent law?
4. Mathematically define probability.
5. Discuss at least four basic properties of probability.
6. Define the following two items:
   i. Cumulative distribution function

    ii. Probability density function
7. Define the expected value of a continuous random variable.
8. Define the following two items:
    i. Laplace transform
    ii. Rayleigh distribution
9. What are the special case distributions of the Weibull and bathtub hazard rate curve distributions?
10. Prove that the sum of Equations 2.57 through 2.59 is equal to unity.

References

1. Eves, H., *An Introduction to the History of Mathematics*, Holt, Rinehart and Winston, New York, 1976.
2. Owen, D. B., editor, *On the History of Statistics and Probability*, Marcel Dekker, New York, 1976.
3. Speigel, M. R., *Probability and Statistics*, McGraw-Hill, New York, 1975.
4. Speigel, M. R., *Statistics*, McGraw-Hill, New York, 1961.
5. Dhillon, B. S., *Reliability, Quality, and Safety for Engineers*, CRC Press, Boca Raton, FL, 2004.
6. Lipschutz, S., *Set Theory*, McGraw-Hill, New York, 1964.
7. Lipschutz, S., *Probability*, McGraw-Hill, New York, 1965.
8. *Fault Tree Handbook*, Report No. NUREG-0492, U.S. Nuclear Regulatory Commission, Washington, DC.
9. Dhillon, B. S., *Computer System Reliability: Safety and Usability*, CRC Press, Boca Raton, FL, 2013.
10. Mann, N. R., Schafer, R. E., and Singpurwalla, N. P., *Methods for Statistical Analysis of Reliability and Life Data*, John Wiley & Sons, New York, 1974.
11. Spiegel, M. R., *Laplace Transforms*, McGraw-Hill, New York, 1965.
12. Oberhettinger, F., and Badic, L., *Tables of Laplace Transforms*, Springer-Verlag, New York, 1973.
13. Patel, J. K., Kapadia, C. H., and Owen, D. H., *Handbook of Statistical Distributions*, Marcel Dekker, New York, 1976.
14. Shooman, M. L., *Probabilistic Reliability: An Engineering Approach*, McGraw-Hill, New York, 1968.
15. Dhillon, B. S., *Reliability Engineering in Systems Design and Operation*, Van Nostrand Reinhold, New York, 1983.
16. Davis, D. J., Analysis of Some Failure Data, *Journal of the American Statistical Association*, June 1952, pp. 113–150.
17. Weibull, W., A Statistical Distribution Function of Wide Applicability, *Journal of Applied Mechanics*, Vol. 18, 1951, pp. 293–297.
18. Dhillon, B. S., Life Distributions, *IEEE Transactions on Reliability*, Vol. 30, 1981, pp. 457–460.
19. Baker, R. D., Nonparametric Estimation of the Renewal Function, *Computers & Operations Research*, Vol. 20, No. 2, 1993, pp. 167–178.
20. Cabana, A., and Cabana, E. M., Goodness-of-Fit to the Exponential Distribution, Focused on Weibull Alternatives, *Communications in Statistics–Simulation and Computation*, Vol. 34, 2005, pp. 711–723.

21. Grane, A., and Fortiana, J., A Directional Test of Exponentiality Based on Maximum Correlations, *Metrika*, Vol. 73, 2011, pp. 255–274.
22. Henze, N., and Meintnis, S. G., Recent and Classical Tests for Exponentiality: A Partial Review with Comparisons, *Metrika*, Vol. 61, 2005, pp. 29–45.
23. Jammalamadaka, S. R., and Taufer, E., Testing Exponentiality by Comparing the Empirical Distribution Function of the Normalized Spacings with That of the Original Data, *Journal of Nonparametric Statistics*, Vol. 15, No. 6, 2003, pp. 719–729.
24. Hollander, M., Laird, G., and Song, K. S., Nonparametric Interference for the Proportionality Function in the Random Censorship Model, *Nonparametric Statistics*, Vol. 15, No. 2, 2003, pp. 151–169.
25. Jammalamadaka, S. R., and Taufer, E., Use of Mean Residual Life in Testing Departures from Exponentiality, *Journal of Nonparametric Statistics*, Vol. 18, No. 3, 2006, pp. 277–292.
26. Kunitz, H., and Pamme, H., The Mixed Gamma Ageing Model in Life Data Analysis, *Statistical Papers*, Vol. 34, 1993, pp. 303–318.
27. Kunitz, H., A New Class of Bathtub-Shaped Hazard Rates and Its Application in Comparison of Two Test-Statistics, *IEEE Transactions on Reliability*, Vol. 38, No. 3, 1989, pp. 351–354.
28. Meintanis, S. G., A Class of Tests for Exponentiality Based on a Continuum of Moment Conditions, *Kybernetika*, Vol. 45, No. 6, 2009, pp. 946–959.
29. Morris, K., and Szynal, D., Goodness-of-Fit Tests Based on Characterizations Involving Moments of Order Statistics, *International Journal of Pure and Applied Mathematics*, Vol. 38, No. 1, 2007, pp. 83–121.
30. Na, M. H., Spline Hazard Rate Estimation Using Censored Data, *Journal of KSIAM*, Vol. 3, No. 2, 1999, pp. 99–106.
31. Morris, K., and Szynal, D., Some U-Statistics in Goodness-of-Fit Tests Derived from Characterizations via Record Values, *International Journal of Pure and Applied Mathematics*, Vol. 46, No. 4, 2008, pp. 339–414.
32. Nam, K. H., and Park, D. H., Failure Rate for Dhillon Model, *Proceedings of the Spring Conference of the Korean Statistical Society*, 1997, pp. 114–118.
33. Nimoto, N., and Zitikis, R., The Atkinson Index, the Moran Statistic, and Testing Exponentiality, *Journal of the Japan Statistical Society*, Vol. 38, No. 2, 2008, pp. 187–205.
34. Nam, K. H., and Chang, S. J., Approximation of the Renewal Function for Hjorth Model and Dhillon Model, *Journal of the Korean Society for Quality Management*, Vol. 34, No. 1, 2006, pp. 34–39.
35. Noughabi, H. A., and Arghami, N. R., Testing Exponentiality Based on Characterizations of the Exponential Distribution, *Journal of Statistical Computation and Simulation*, Vol. 81, No. 11, 2011, pp. 1641–1651.
36. Szynal, D., Goodness-of-Fit Tests Derived from Characterizations of Continuous Distributions, Stability in Probability, *Banach Center Publications*, Vol. 90, 2010, pp. 203–223.
37. Szynal, D., and Wolynski, W., Goodness-of-Fit Tests for Exponentiality and Rayleigh Distribution, *International Journal of Pure and Applied Mathematics*, Vol. 78, No. 5, 2012, pp. 751–772.

38. Nam, K. H., and Park, D. H., A Study on Trend Changes for Certain Parametric Families, *Journal of the Korean Society for Quality Management*, Vol. 23, No. 3, 1995, pp. 93–101.
39. Srivastava, A. K., Validation Analysis of Dhillon Model on Different Real Data Sets for Reliability Modelling, *International Journal of Advance Foundation and Research in Computer (IJAFRC)*, Vol. 1, No. 9, 2014, pp. 18–31.
40. Srivastava, A. K., and Kumar, V., A Study of Several Issues of Reliability Modelling for a Real Dataset Using Different Software Reliability Models, *International Journal of Emerging Technology and Advanced Engineering*, Vol. 5, No. 12, 2015, pp. 49–57.
41. Lim, Y. B. et al., Literature Review on the Statistical Methods in KSQM for 50 years, *Journal of the Korean Society for Quality Management*, Vol. 44, No. 2, 2016, pp. 221–244.
42. Zendehdel, J. et al., Testing Exponentiality for Imprecise Data and Its Application, Soft Computing: A Fusion of Foundations, *Methodologies and Applications*, Vol. 21, March 2017, pp. 1–12.
43. Sathish, K., and Geetha, T., A mathematical model using for the impact of corticotrophin releasing hormone on gastrointestinal motility and acth hormone in normal controls and patients with irritable bowel syndrome, *International Journal of Engineering, Science and Mathematics*, Vol. 7, No. 2, 2018, pp. 96–103.

chapter three

Reliability basics and human factor basics for usability

3.1 Introduction

The history of the reliability field may be traced to the early years of 1930s, when probability concepts were applied to problems in the area of electric power generation [1–3]. However, its real beginning is generally regarded as World War II, when German scientists applied basic reliability concepts for improving the performance of their V1 and V2 rockets [4]. Today, the reliability field has become a well-developed discipline and has branched out into a number of specialized areas, such as software reliability, power system reliability, mechanical reliability, and human reliability and error [3,4].

The history of the field of human factors may be traced to 1898, when Frederick W. Taylor conducted various studies for determining the most effective design of shovels [5]. In 1911, Frank B. Gilbreth invented a scaffold that allowed bricklayers to carry out their activities at the most appropriate levels at all times. From 1911–1945, many new developments took place in the area of human factors, and in 1945, it became a recognized discipline. In 1957, the Human Factors Society of America was incorporated, and in 1992, it became known as the Human Factors and Ergonomics Society of America. Needless to say, human factors is a well-developed field nowadays. Additional information on the history of reliability and the human factor field is available in References 4 and 6, respectively.

This chapter presents the reliability basics and human factor basics for usability, which are considered vital to understanding subsequent chapters of this book.

3.2 Bathtub hazard rate curve

The bathtub hazard rate curve (shown in Figure 3.1) is usually used to describe the failure rate of engineering systems. The curve is divided into three parts: the burn-in period, the useful-life period, and the wear-out period.

During the burn-in period, the engineering system/item hazard rate decreases with time t, and some of the reasons for failures during this

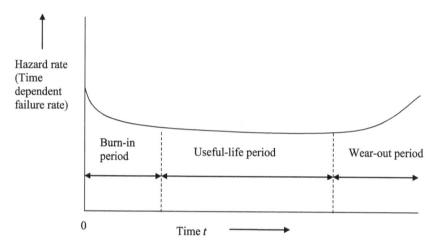

Figure 3.1 Bathtub hazard rate curve.

period include substandard materials and workmanship, poor manufacturing methods, inadequate debugging, poor processes, human error, and poor quality control [4,7]. Note that three other terms used for this decreasing hazard rate region (i.e., the burn-in period) are *infant mortality region, break-in region,* and *debugging region.*

During the useful life period, the system/item hazard rate remains constant with time. Some of the reasons for failures during this period are higher random stress than expected, low safety factors, undetectable defects, abuse, human errors, and natural failures.

Finally, during the wear-out period, the system/item hazard rate increases with time *t*. Some of the reasons for failures during this period are wear due to friction, corrosion, and creep; poor maintenance, incorrect overhaul practices, wear due to aging, and short designed-in life of the system/item under consideration.

The following equation can be used to represent the bathtub hazard rate curve mathematically [8]:

$$\lambda(t) = \theta\alpha(\theta t)^{\alpha-1}e^{(\theta t)^{\alpha}}, \tag{3.1}$$

where $\lambda(t)$ is the hazard rate (time-dependent failure rate), *t* is time, α is the shape parameter, and θ is the scale parameter.

At $\alpha = 0.5$, Equation 3.1 gives the shape of the bathtub hazard rate curve shown in Figure 3.1.

3.3 General reliability analysis-related formulas

There are a number of general formulas for performing various types of reliability analysis. Four of these formulas are presented in the next subsections.

3.3.1 Failure (or probability) density function

The failure (or probability) density function is expressed by Reference 4 as follows:

$$f(t) = -\frac{dR(t)}{dt}, \tag{3.2}$$

where $f(t)$ is the failure (or probability) density function, t is time, and $R(t)$ is the system/item reliability at time t.

EXAMPLE 3.1

Assume that the reliability of an engineering system is expressed by the following equation:

$$R_{es}(t) = e^{-\lambda_{es}t}, \tag{3.3}$$

where λ_{es} is the engineering system constant failure rate, t is time, and $R_{es}(t)$ is the engineering system reliability at time t.

Obtain an expression for the failure (or probability) density function of the engineering system by using Equation 3.2.

By substituting Equation 3.3 into Equation 3.2, we get

$$f(t) = -\frac{de^{-\lambda_{es}t}}{dt}$$
$$= \lambda_{es}e^{-\lambda_{es}t}. \tag{3.4}$$

Thus, Equation 3.4 is the expression for the failure (or probability) density function of the engineering system.

3.3.2 Hazard rate function

The hazard rate function is defined by

$$\lambda(t) = \frac{f(t)}{R(t)}, \tag{3.5}$$

where $\lambda(t)$ is the system/item hazard rate (i.e., time-dependent failure rate).

Inserting Equation 3.2 into Equation 3.5 yields

$$\lambda(t) = -\frac{1}{R(t)} \cdot \frac{dR(t)}{dt}. \tag{3.6}$$

EXAMPLE 3.2

Obtain an expression for the hazard rate of the engineering system with the aid of Equations 3.3 and 3.6 and comment on the result.

By inserting Equation 3.3 into Equation 3.6, we obtain

$$\lambda(t) = -\frac{1}{e^{-\lambda_{es}t}} \cdot \frac{de^{-\lambda_{es}t}}{dt}$$

$$= \lambda_{es}. \tag{3.7}$$

Thus, the hazard rate of the engineering system is given by Equation 3.7, and the right side of this equation is not a function of time t. Needless to say, λ_{es} is usually referred to as the constant failure rate of a system/item (in this case, of the engineering system) because it does not depend on time t.

3.3.3 General reliability function

The general reliability function can be obtained by using Equation 3.6 as follows:

$$-\lambda(t)dt = \frac{1}{R(t)}dR(t). \tag{3.8}$$

By integrating both sides of Equation 3.8 over the time interval $[0,t]$, we get

$$-\int_0^t \lambda(t)dt = \int_1^{R(t)} \frac{1}{R(t)}dR(t). \tag{3.9}$$

At $t = 0$, $R(t) = 1$.

By evaluating the right side of Equation 3.9 and rearranging, we get

$$\ln R(t) = -\int_0^t \lambda(t)dt. \tag{3.10}$$

Thus, from Equation 3.10, we get

$$R(t) = e^{-\int_0^t \lambda(t)dt}. \tag{3.11}$$

Equation 3.11 is the general expression for the reliability function. Thus, it can be used to obtain the reliability function of an item/system when its times to failure follow any time-continuous probability distribution (e.g., exponential, Rayleigh, and Weibull).

EXAMPLE 3.3

Assume that the times to failure of an engineering system are exponentially distributed and that the constant failure rate is 0.004 failures/h. Calculate the reliability of the engineering system for a 20-h mission.

By inserting the specified data values into Equation 3.11, we obtain

$$R(20) = e^{-\int_0^{20} (0.004)dt}$$

$$= e^{-(0.004)(20)}$$

$$= 0.9231.$$

Thus, the reliability of the engineering system is 0.9231. In other words, there is a 92.31% chance that the engineering system will not fail during the given time period.

3.3.4   Mean time to failure

The mean time to failure of an item/system can be obtained by using any of the following three formulas [4,9]:

$$\text{MTTF} = \int_0^\infty R(t)dt \tag{3.12}$$

or

$$\text{MTTF} = \lim_{s \to 0} R(s) \tag{3.13}$$

or

$$\text{MTTF} = E(t) = \int_0^\infty tf(t)dt, \tag{3.14}$$

where MTTF is the mean time to failure, $E(t)$ is the expected value, s is the Laplace transform variable, and $R(s)$ is the Laplace transform of the reliability function $R(t)$.

EXAMPLE 3.4

Prove by using Equation 3.3 that Equations 3.12 and 3.13 yield the same result for the engineering system's mean time to failure.

By inserting Equation 3.3 into Equation 3.12, we get

$$\text{MTTF}_{es} = \int_0^\infty e^{-\lambda_{es}t}dt \tag{3.15}$$

$$= \frac{1}{\lambda_{es}},$$

where MTTF_{ns} is the engineering system's mean time to failure.

By taking the Laplace transform of Equation 3.3, we obtain

$$R_{es}(s) = \int_0^\infty e^{-st}e^{-\lambda_{es}t}dt \tag{3.16}$$

$$= \frac{1}{s + \lambda_{es}},$$

where $R_{es}(s)$ is the Laplace transform of the engineering system reliability function $R_{es}(t)$.

By substituting Equation 3.16 into Equation 3.13, we obtain

$$\text{MTTF}_{es} = \lim_{s \to 0} \frac{1}{(s + \lambda_{es})} \tag{3.17}$$

$$= \frac{1}{\lambda_{es}}.$$

Equations 3.15 and 3.17 are identical, which proves that Equations 3.12 and 3.13 yield the same result for the engineering system's mean time to failure.

3.4　Reliability networks

An engineering system can form various configurations in performing reliability analysis. This section is concerned with the reliability evaluation of such commonly occurring configurations or networks.

3.4.1　Series network

The series network is the simplest reliability network or configuration, and its block diagram is shown in Figure 3.2. The diagram denotes an n-unit system, and each block in the diagram represents a unit. For the successful operation of the series system, all n units must operate normally. In other

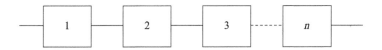

Figure 3.2   Block diagram of a series system with *n* units.

words, if any one of the *n* units malfunctions or fails, the series system as a whole fails.

With the series system, shown in Figure 3.2, reliability is expressed by

$$R_{ss} = P(E_1 E_2 E_3 \ldots E_n), \tag{3.18}$$

where R_{ss} is the series system reliability; E_j is the successful operation (i.e., success event) of unit j, for $j = 1,2,3, \ldots ,n$; and $P(E_1 E_2 E_3 \ldots E_n)$ is the occurrence probability of events $E_1, E_2, E_3, \ldots E_n$.

For independently failing units, Equation 3.18 becomes

$$R_{ss} = P(E_1)P(E_2)P(E_3) \ldots P(E_n), \tag{3.19}$$

where $P(E_j)$ is the occurrence probability of event E_j, for $j = 1,2,3,\ldots ,n$.

If we let $R_j = P(E_j)$, for $j = 1,2,3, \ldots ,n$, Equation 3.19 becomes

$$R_{ss} = R_1 R_2 R_3 \ldots R_n$$

$$= \prod_{j=1}^{n} R_j, \tag{3.20}$$

where R_j is the unit j reliability, for $j = 1,2,3, \ldots ,n$.

For the constant failure rate λ_j of unit j from Equation 3.11 (i.e., $\lambda_j(t) = \lambda_j$), we get

$$R_j(t) = e^{-\lambda_j t}, \tag{3.21}$$

where $R_j(t)$ is the unit j reliability at time t.

By inserting Equation 3.21 into Equation 3.20, we obtain

$$R_{ss}(t) = e^{-\sum_{j=1}^{n} \lambda_j t}, \tag{3.22}$$

where $R_{ss}(t)$ is the series system reliability at time t.

Using Equation 3.22 in Equation 3.12, we obtain the following expression for the series system's mean time to failure:

$$\mathrm{MTTF}_{ss} = \int_0^{\infty} e^{-\sum_{j=1}^{n} \lambda_j t} dt$$

$$= \frac{1}{\sum_{j=1}^{n} \lambda_j}, \tag{3.23}$$

where MTTF_{ss} is the series system's mean time to failure.

By inserting Equation 3.22 into Equation 3.6, we obtain the following expression for the series system's hazard rate:

$$\lambda_{ss}(t) = -\frac{1}{e^{-\sum_{j=1}^{n} \lambda_j t}} \left[-\sum_{j=1}^{n} \lambda_j \right] e^{-\sum_{j=1}^{n} \lambda_j t}$$

$$= \sum_{j=1}^{n} \lambda_j,$$

(3.24)

where $\lambda_{ss}(t)$ is the series system's hazard rate.

Note that the right side of Equation 3.24 is independent of time t. Thus, the left side of this equation is simply λ_{ss}, the failure rate of the series system. It simply means that whenever we add up the failure rates of units/items, we automatically assume that these units/items form a series configuration/network—a worst-case design scenario with regard to reliability.

EXAMPLE 3.5

Assume that an engineering system is composed of two independent and identical subsystems, and that the constant failure rate of a subsystem is 0.0005 failures/h. Both subsystems must work normally in order for the engineering system to operate successfully. Calculate the following:

- The engineering system's reliability for a 100-h mission
- The engineering system's mean time to failure
- The engineering system's failure rate

In this case, the engineering system's subsystems form a series configuration. Thus, by inserting the given data values into Equation 3.22, we obtain

$$R_{ss}(100) = e^{-(0.0005)(2)(100)}$$
$$= 0.9048.$$

Substituting the specified data values into Equation 3.23 yields

$$\text{MTTF}_{ss} = \frac{1}{2(0.0005)}$$
$$= 1{,}000 \text{ h}.$$

Using the specified data values in Equation 3.24 yields

$$\lambda_{ss} = 2(0.0005)$$
$$= 0.001 \text{ failures/h}$$

Thus, the engineering system reliability, mean time to failure, and failure rate are 0.9048, 1,000 h, and 0.001 failures/h, respectively.

3.4.2   Parallel network

In this case, a parallel network is composed of n simultaneously operating units or items, and for the successful operation of the system, at least one of these units or items must operate normally. The n-unit parallel system or network is shown in Figure 3.3, and each block in the diagram represents a unit.

The probability of failure of the parallel system shown in Figure 3.3 is expressed by

$$F_{ps} = P(\bar{E}_1 \bar{E}_2 \bar{E}_3 \dots \bar{E}_n), \qquad (3.25)$$

where F_{ps} is the probability of failure of the parallel system; $\bar{E}_j$ is the failure (i.e., failure event) of unit j for $j = 1,2,3, \dots ,n$; and $P(\bar{E}_1 \bar{E}_2 \bar{E}_3 \dots \bar{E}_n)$ is the probability of occurrence of events $\bar{E}_1, \bar{E}_2, \bar{E}_3, \dots ,$ and $\bar{E}_n$.

For independently failing parallel units, Equation 3.25 becomes

$$F_{ps} = P(\bar{E}_1)P(\bar{E}_2)P(\bar{E}_3) \dots P(\bar{E}_n), \qquad (3.26)$$

where $P(\bar{E}_j)$ is the occurrence probability of failure event $\bar{E}_j$, for $j = 1,2,3, \dots ,n$.

If we let $F_j = P(\bar{E}_j)$, for $j = 1,2,3, \dots ,n$, then Equation 3.26 becomes

$$F_{ps} = F_1 F_2 F_3 \dots F_n, \qquad (3.27)$$

where F_j is the failure probability of unit j, for $j = 1,2,3, \dots ,n$.

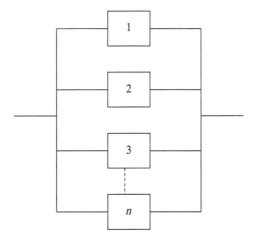

Figure 3.3   Block diagram of a parallel system/network with n units.

By subtracting Equation 3.27 from unity, we get

$$R_{ps} = 1 - F_{ps}$$
$$= 1 - F_1 F_2 F_3 \ldots F_n, \tag{3.28}$$

where R_{ps} is the reliability of the parallel system or network.

For the constant failure rate λ_j of unit j, subtracting Equation 3.21 from unity and then substituting it into Equation 3.28 obtains

$$R_{ps}(t) = 1 - (1 - e^{-\lambda_1 t})(1 - e^{-\lambda_2 t})(1 - e^{-\lambda_3 t}) \ldots (1 - e^{-\lambda_n t}), \tag{3.29}$$

where $R_{ps}(t)$ is the reliability of the parallel system or network at time t.

For identical units, inserting Equation 3.29 into Equation 3.12 obtains the following expression for the parallel system's mean time to failure:

$$\text{MTTF}_{ps} = \int_0^\infty [1 - (1 - e^{-\lambda t})^n] dt$$
$$= \frac{1}{\lambda} \sum_{j=1}^n \frac{1}{j}, \tag{3.30}$$

where MTTF_{ps} is the mean time to failure of the parallel system or network and λ is the unit constant failure rate.

EXAMPLE 3.6

Assume that an engineering system is composed of two independent, identical, and active units. At least one of these units must operate normally in order for the engineering system to operate successfully.

Calculate the reliability of the engineering system if each unit's failure probability is 0.2.

By substituting the specified data values into Equation 3.28, we obtain

$$R_{ps} = 1 - (0.2)(0.2)$$
$$= 0.96.$$

Thus, the reliability of the engineering system is 0.96.

EXAMPLE 3.7

Assume that an engineering system has three identical and independent units in parallel. The constant failure rate of a unit is 0.008 failures/h. Calculate the engineering system's mean time to failure.

By inserting the given data values into Equation 3.30, we obtain

$$\text{MTTM}_{ps} = \frac{1}{(0.008)}\left[1+\frac{1}{2}+\frac{1}{3}\right]$$

$$= 229.16 \text{ h.}$$

Thus, the engineering system's mean time to failure is 229.16 h.

3.4.3 k-out-of-n network

The k-out-of-n network is another form of redundancy in which, for the successful operation of the system, at least k units out of a total of n units must operate normally. The block diagram of a k-out-of-n unit system or network is shown in Figure 3.4. Each block in the diagram denotes a unit. Note that the series and parallel networks are special cases of this network for $k = n$ and $k = 1$, respectively.

With the aid of binomial distribution, for independent and identical units, we get the following expression of the reliability of k-out-of-n unit system or network shown in Figure 3.4:

$$R_{k/n} = \sum_{j=k}^{n}\binom{n}{j}R^j(1-R)^{n-j}, \tag{3.31}$$

where

$$\binom{n}{j} = \frac{n!}{(n-j)!j!}, \tag{3.32}$$

R is the unit reliability, and $R_{k/n}$ is the reliability of the k-out-of-n network or system.

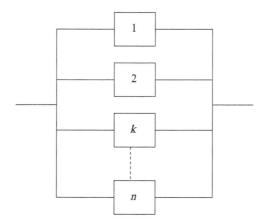

Figure 3.4 k-out-of-n unit system/network block diagram.

For constant failure rates of the identical units, using Equations 3.11 and 3.31, we obtain

$$R_{k/n}(t) = \sum_{j=k}^{n} \binom{n}{j} e^{-j\lambda t}(1 - e^{-\lambda t})^{n-j}, \tag{3.33}$$

where λ is the unit constant failure rate and $R_{k/n}(t)$ is the reliability of the k-out-of-n network or system at time t.

By inserting Equation 3.33 into Equation 3.12, we get

$$\text{MTTF}_{k/n} = \int_{0}^{\infty} \left[\sum_{j=k}^{n} \binom{n}{j} e^{-j\lambda t}(1 - e^{-\lambda t})^{n-j} \right] dt$$

$$= \frac{1}{\lambda} \sum_{j=k}^{n} \frac{1}{j}, \tag{3.34}$$

where $\text{MTTF}_{k/n}$ is the mean time to failure of the k-out-of-n network or system.

EXAMPLE 3.8

Assume that an engineering system has four identical, independent, and active units in parallel. At least three units must operate normally for the successful operation of the engineering system. Calculate the engineering system's mean time to failure if the unit constant failure rate is 0.0004 failures/h.

By substituting the given data values into Equation 3.34, we obtain

$$\text{MTTF}_{3/4} = \frac{1}{(0.0004)} \sum_{j=3}^{4} \frac{1}{j}$$

$$= 1{,}458.33 \text{ h.}$$

Thus, the engineering system's mean time to failure is 1,458.33 h.

3.4.4   Standby system

The standby system is another network or system in which only one unit operates and m units are kept in standby mode. The entire system contains $(m + 1)$ units, and as soon as the operating unit fails, the switching mechanism detects the failure and turns on one of the standby units.

A standby system's block diagram representing 1 operating unit and m standby units is shown in Figure 3.5. Each block in the diagram denotes a unit.

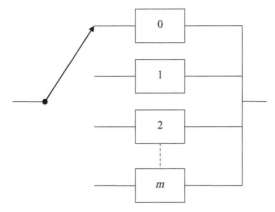

Figure 3.5   Block diagram of a standby system with 1 operating unit and *m* standby units.

With the aid of the block diagram for independent and identical units shown in Figure 3.5, the perfect switching mechanism and standby units, and the time-dependent unit failure rate, we write the following equation to express the standby system's reliability [10]:

$$R_{ss}(t) = \sum_{j=0}^{m} \left[\left[\int_0^t \lambda(t)dt \right]^j e^{-\int_0^t \lambda(t)dt} \right] / j!, \qquad (3.35)$$

where $R_{ss}(t)$ is the standby system reliability at time t and $\lambda(t)$ is the unit's time-dependent failure rate or hazard rate.

For the constant unit failure rate (i.e., $\lambda(t) = \lambda$), Equation 3.35 yields

$$R_{ss}(t) = \sum_{j=0}^{m} [(\lambda t)^j e^{-\lambda t}] / j!, \qquad (3.36)$$

where λ is the unit constant failure rate.

By inserting Equation 3.36 into Equation 3.12, we get

$$\mathrm{MTTF}_{ss} = \int_0^\infty \left[\frac{\sum_{j=0}^{m} \{ (\lambda t)^j e^{-\lambda t} \}}{j!} \right] dt \qquad (3.37)$$

$$= \frac{m+1}{\lambda},$$

where MTTF_{ss} is the standby system's mean time to failure.

EXAMPLE 3.9

Assume that an engineering standby system contains two identical, independent units (one operating and the other on standby). The unit constant failure rate is 0.008 failures/h. Calculate the standby system reliability for a 200-h mission and mean time to failure, assuming that the switching mechanism is perfect and the standby unit remains as good as new in its standby mode.

By substituting the specified data values into Equation 3.36, we get

$$R_{ss}(200) = \frac{\sum_{j=0}^{1}\left[[(0.008)(200)]^j e^{-(0.008)(200)}\right]}{j!}$$

$$= 0.5249.$$

Similarly, by inserting the given data values into Equation 3.37, we obtain

$$\text{MTTF}_{ss} = \frac{2}{0.008}$$

$$= 250 \text{ h}.$$

Thus, the standby system's reliability and mean time to failure are 0.5249 and 250 h, respectively.

3.4.5 Bridge network

Sometimes units in engineering systems may form a bridge network or configuration, as shown in Figure 3.6. Each block in the figure denotes a unit, and all five units are labeled with numerals.

For independent units, the reliability of the network in Figure 3.6 is expressed by Reference 11 as follows:

$$R_{bcn} = 2R_1R_2R_3R_4R_5 + R_1R_3R_5 + R_2R_3R_4 + R_2R_5 + R_1R_4$$
$$- R_1R_2R_3R_4 - R_1R_2R_3R_5 - R_2R_3R_4R_5 - R_1R_2R_4R_5 - R_3R_4R_5R_1.$$

$$(3.38)$$

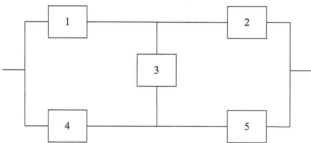

Figure 3.6 A bridge configuration or network of five dissimilar units.

where R_{bcn} is the reliability of the bridge configuration or network and R_j is the reliability of the jth unit, for $j = 1,2,3,4,5$.

For identical units, Equation 3.38 becomes

$$R_{bcn} = 2R^5 - 5R^4 + 2R^3 + 2R_2, \tag{3.39}$$

where R is the unit reliability.

For constant unit failure rate, using Equations 3.11 and 3.39, we obtain

$$R_{bcn}(t) = 2e^{-5\lambda t} - 5e^{-4\lambda t} + 2e^{-3\lambda t} + 2e^{-2\lambda t}, \tag{3.40}$$

where $R_{bcn}(t)$ is the reliability of the bridge configuration or network at time t and λ is the unit constant failure rate.

By substituting Equation 3.40 into Equation 3.12, we get

$$\begin{aligned}
\text{MTTF}_{bcn} &= \int_0^\infty (2e^{-5\lambda t} - 5e^{-4\lambda t} + 2e^{-3\lambda t} + 2e^{-2\lambda t})dt \\
&= \frac{49}{60\lambda}.
\end{aligned} \tag{3.41}$$

where MTTF_{bcn} is the mean time to failure of the bridge configuration or network.

EXAMPLE 3.10

Assume that an engineering system has five identical and independent units forming a bridge configuration or network. The constant failure rate of each unit is 0.0004 failures/h.

Calculate the reliability of the bridge configuration or network for a 100-h mission and the mean time to failure.

By inserting the specified data values into Equation 3.40, we obtain

$$R_{bcn}(100) = 2e^{-5(0.0004)(100)} - 5e^{-4(0.0004)(100)} + 2e^{-3(0.0004)(100)}$$
$$+ 2e^{-2(0.0004)(100)} = 0.9968.$$

Similarly, by substituting the given data values into Equation 3.41, we obtain

$$\text{MTTF}_{bcn} = \frac{49}{60(0.0004)}$$
$$= 2,041.66\,\text{h}.$$

Thus, the reliability of the bridge configuration or network and the mean time to failure are 0.9968 and 2,041.66 h, respectively.

3.5 Human factor objectives and typical human behaviors

There are many human factor objectives, and they may be grouped under the following four categories [12,13]:

1. *Category I*: This includes fundamental operational objectives: to increase safety, improve system performance, and reduce the occurrence of errors.
2. *Category II*: This includes objectives that, directly or indirectly, have a bearing on reliability, maintainability, availability, and integrated logistic support: to increase reliability, improve maintainability, reduce personnel requirements, and reduce training requirements.
3. *Category III*: This includes objectives that, directly or indirectly, affect users and operators: to increase user acceptance, increase ease of use, reduce boredom and monotony, improve the working environment, increase aesthetic appearance, increase human comfort, and reduce fatigue and physical stress.
4. *Category IV*: This includes miscellaneous objectives, such as to reduce the loss of time and equipment and to increase production economy.

Over the years, researchers in the area of human factors have conducted extensive research on predicting human behavior. They have fully highlighted a great many typical human behaviors. Some of these behaviors, along with the proposed design-related measures in parentheses and italic, are presented here [13,14]:

- People have a tendency to hurry. (*Design products/items in such a way that effectively takes into consideration the element of hurry by humans.*)
- People have become very accustomed to specific meanings of color. (*Strictly observe current color-coding standards during the design process.*)
- People expect electrically powered switches to move upward, to the right, etc., to activate power. (*Ensure that such devices are properly designed according to human expectations.*)
- Generally, people know very little about their physical shortcomings. (*First, learn effectively about all human limitations and then develop the design accordingly.*)
- People tend to regard all manufactured items as being safe. (*Design products/items in such a manner so that they cannot be used improperly. If this is not possible, then design in an appropriate mechanism to make all potential users clearly aware of all possible hazards.*)

- People expect that faucet and valve handles rotate counterclockwise for increasing the flow of a liquid, gas, or steam. (*Ensure that devices are designed to conform to human expectations.*)
- During loss of balance, people instinctively reach for and grab the closest object/item. (*Develop the design of the product/item in such a manner that it properly incorporates satisfactory emergency supports.*)
- People's attention is drawn to items such as loud noises, flashing lights, bright lights, and bright and vivid colors. (*Ensure that stimuli of adequate intensity are properly designed for when attention needs to be attracted.*)
- Generally, people use their hands first for testing or exploring. (*First, pay special attention to the handling aspect during the product/item design process. Otherwise, recommend clearly and strongly that the product/item use requires a device supplied to eliminate the need to use the hands.*)
- People get easily confused with unfamiliar products and items. (*Avoid designing products/items that are totally unfamiliar to potential users.*)

3.6 Types of human-machine systems and human-machine comparisons

There are basically three types of human-machine systems, as shown in Figure 3.7 [13,15]. The types shown in the figure are automated systems, semiautomatic/mechanical systems, and manual systems.

Automated systems perform all operational functions, including sensing, information processing and decision-making, and action. These systems are required to be programmed for taking all appropriate measures to respond to all contingencies as they take place. Three specific human-related functions associated with automated systems are monitoring, programming, and maintenance.

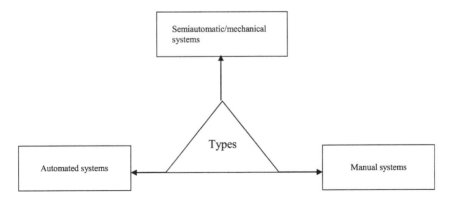

Figure 3.7 Types of human-machine systems.

Semiautomatic/mechanical systems are composed of well-integrated physical parts, such as various types of powered machine tools, and are generally designed for conducting their functions with minor variation. In these systems, typically the machine provides power, and the involved person essentially takes care of the controlling aspect of the operation.

Finally, manual systems are composed of hand tools and other aids that are coupled together by the person controlling the operation. Generally, a craft worker is the person who is involved in the operation. Nonetheless, operators of these systems use their own energy as a power source, and they also transmit and receive information from these systems during use.

During the design of engineering systems, sometimes decisions may have to be made whether to allocate certain functions to humans or to machines. In such situations, an effective understanding of humans' and machines' capabilities and limitations is absolutely necessary; otherwise, the correct decisions may not be made. Table 3.1 presents a number of comparisons of humans' and machines' capabilities and limitations [13,14].

3.7   Human body dimensions and sensory capacities

Information on human body dimensions is very important to designers so they can allocate the proper workspace and tasks to involved individuals because most engineering products and items are operated and maintained by humans. Generally, requirements associated with the human body are outlined in the specifications of product design. Two basic sources for obtaining information on body dimensions are experiments and anthropometric surveys.

Experiments simulate the conditions in question to collect the required data. First, the relevant conditions are simulated by experiments; and then the required data are collected. With anthropometric surveys, measurements of a sample of the population are taken. Generally, the findings of these measurements are presented in the form of means, ranges, and percentiles.

The body dimensions and measurements are classified under the following two groups:

1. *Static*: In this case, measurements include items ranging from dimensions of the body's largest aspects to measurements of the distance between the pupils of the eyes.
2. *Dynamic*: In this case, measurements generally vary based on body movements and include those taken with subjects in various working positions, as well as arm and leg reaches.

Table 3.2 presents selected body dimensions (in inches) of females and males for U.S. adults ranging in age from 18–79 years [13–16].

Table 3.1 A comparison of humans' and machines' capabilities and limitations

No.	Human capability/limitation	Machine capability/limitation
1	Excellent memory	Very expensive to have same memory capability as humans
2	Subject to social environment	Free of social environment
3	Extremely limited short-term memory for factual matters	Short-term memory can be expanded to any desirable and affordable level
4	A high degree of tolerance to factors such as vagueness, ambiguity, and uncertainty	Limited in tolerance to such factors
5	Quite capable in performing time-contingency analyses and predicting events in unfamiliar environments	Very poor at this aspect
6	Limited channel capacity	Channel capacity can be expanded to satisfy the need
7	Very flexible with respect to task performance	Relatively inflexible
8	Subject to degradation in performance because of fatigue and boredom	Subject to degradation in performance because of wear or lack of calibration
9	Prone to stress as the result of interpersonal or other related problems	Independent of such problems
10	Highly capable of making inductive decisions in novel situations	Little or no induction capability
11	Relatively easy maintenance	Increase in complexity leads to serious maintenance problems
12	Prone to factors such as motion sickness, coriolis effects, and disorientation	Free from such factors
13	Optimum strategy may not be followed all the time	Designed strategy is executed all the time
14	Quite adversely affected by high g-forces	Independent of g-forces
15	Quite capable to interpret an input signal even in distractive, noisy, and similar conditions	Usually performs well only under ideal environments (i.e., noise-free, clean, etc.)
16	Subject to psychological, ecological, and physiological needs	Subject to ecological needs only
17	Performance efficiency is affected by the anxiety factor	Performance efficiency is unaffected by anxiety
18	Quite capable of performing under transient overload (performance degrades gracefully)	Operation stops under overload conditions and generally fails at once
19	Quite unsuitable to perform tasks such as data coding, amplification, or transformation	Extremely useful to perform such tasks

Table 3.2 Selected body dimensions of females and males

No.	Body Feature	Female (measured in inches)		Male (measured in inches)	
		5th (percentile)	95th (percentile)	5th (percentile)	95th (percentile)
1	Height, standing	59	67.1	63.8	72.8
2	Seated eye height	27.4	31.0	28.7	33.5
3	Height, sitting erect	30.9	35.7	33.2	38.0
4	Knee height	17.9	21.5	19.3	23.4
5	Seat breadth	12.3	17.1	12.2	15.9

Note: U.S. adults, ranging in age from 18–79 years.

3.7.1 Sensory capacities

Humans possess many useful senses: sight, touch, hearing, taste, and smell. A good understanding of these sensors can be very useful for reducing various types of usability-related problems. Three of them (i.e., sight, touch, and hearing) are described in the next subsections [13,17].

3.7.1.1 Sight

The sense of sight is stimulated by electromagnetic radiation of certain wavelengths, frequently referred to as the *electromagnetic spectrum*. The parts of the spectrum, as seen by the human eye, appear to vary in brightness. According to studies performed over the years, in daylight, the eyes of humans are most sensitive to greenish-yellow light with a wavelength of around 5,500 Å [17]. Moreover, the eyes see differently from different angles.

Some of the important sight-associated guidelines are as follows:

- Select colors in such a way so that color-weak individuals do not get confused.
- Aim to use red filters with wavelengths more than 6,500 Å.
- Avoid relying on color as much as possible (where critical tasks may be conducted by fatigued individuals).

3.7.1.2 Touch

This is quite closely related to humans' ability to interpret visual and auditory stimuli. The sensory cues received by the skin and muscles can be used for sending messages to the brain. In turn, this helps to relieve a part of the load from ears and eyes. This human quality can be utilized quite successfully in various engineering usability areas. For example, in situations when the user of a product or item is expected to rely totally on his or her touch, different shapes of knob could be considered for use.

Finally, the use of touch in various technical areas is not new; it has been utilized for many centuries by artisans for detecting surface irregularities and roughness in their work. In fact, past experience over the years clearly indicate that the detection accuracy of surface irregularities improves dramatically when the involved individual moves an intermediate piece of paper or thin cloth over the surface of the object under consideration instead of just bare fingers [18].

3.7.1.3 Noise (hearing)

The hearing sense may be defined simply as sounds that lack coherence, and reaction of the human to noise extends well beyond the auditory systems (i.e., to feelings such as boredom, fatigue, well-being, or irritability). Excessive noise can result in various types of problems, including loss of hearing if exposed for long periods, adverse effects on tasks requiring a high degree of muscular coordination and precision or intense concentration, and reduction in worker efficiency.

The human ear can detect sounds with frequencies ranging from 20–20,000 Hz and is most sensitive to frequencies in the range of 600–900 Hz. Finally, note that humans exposed to noise with frequencies between 4,000 and 6,000 Hz for long periods can suffer major loss of hearing [17,19].

3.8 Effects of vibrations on humans

Vibration may simply be described as the alternating motion of a body or surface with regard to a reference point. Generally, engineering products (e.g., machines, appliances, and vehicles) vibrate. Moreover, vibrations may be transmitted to users and operators of products by direct mechanical path or through various flanking paths, and the vibrating objects and items can make either the whole body or part of it vibrate.

The human body's physical responses to vibration are the results of complex and sophisticated interactions between body masses, couplings, elasticities, and damping in the low-frequency range (i.e., up to 50 Hz). Human body parts such as the hands, shoulders, head, and eyeballs have certain resonant frequencies that amplify the motions transmitted to them. However, these frequencies can be altered through measures such as handholds, cushioned seats, foot restraints, and mountings.

Vibrations can affect human performance in the following ways [12,13]:

- By blurring visual images of dials and indicators
- By modifying perception
- By affecting control movements

Table 3.3 Six selected activities affected by various vibration frequencies

No.	Activity description	Frequency range (Hz)
1	Reading text	1–50
2	Head movements	6–8
3	Speech	1–20
4	Reading instruments	6–11
5	Tracking	1–30
6	Tactile sensing	30–300

Furthermore, note that excessive vibration can be quite painful, unpleasant, and hazardous. Six important selected activities affected by various vibration frequencies are presented in Table 3.3 [12,13].

3.9 Glare and glare reduction and effective illumination levels

Past experience over the years indicates that glare could be an important factor in the effectiveness of the usability of engineering products. It is produced by brightness within the field of vision that is significantly more than the luminance to which the human eye is accustomed to, causing a loss in visual performance and visibility, discomfort, or annoyance.

There are basically two types of glare. These are direct glare and specular or reflected glare [15]. Direct glare is generated by light sources in the field of view, while specular or reflected glare is generated by highly bright reflections from glossy or polished surfaces that are pointed toward an individual, such as the reflection of overhead light of a computer terminal's cathode-ray-tube (CRT) screen [15,20].

Variations in the level of glare can be classified into three types, as shown in Figure 3.8.

Disability glare decreases visibility and visual performance and generally is accompanied by a certain degree of discomfort. The blinding glare is extremely intense, and even after its removal for an appreciable length of time, no object can be seen.

Finally, discomfort glare causes distress but does not necessarily interfere with visibility or visual-related performance. The calculation of discomfort glare ratings for specific lighting layouts takes into consideration various situational-related factors that, directly or indirectly, affect visual comfort, including room surface reflectance, room size and shape, number and location of luminaries, illumination level, and luminaire size, type, and light distribution.

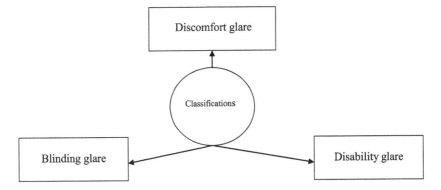

Figure 3.8 Classification of variations in the level of glare.

Over the years, various ways and means have been proposed for reducing different types of glare. Three of these (i.e., ways and means) are described here [13,15]:

1. *Direct glare from luminaires*: Some of the important measures for reducing the direct glare from luminaires are as follows:
   a. Increase the luminance of the area around the glare source so that the luminance (brightness) ratio is lowest.
   b. Place all luminaires as far from the line of sight as possible.
   c. Use hoods, visors, and light shields when it is impossible to lower glare sources.
   d. Lower the light sources' luminance.
   e. Choose luminaires with low discomfort glare ratings.
2. *Direct glare from windows*: Four of the important measures for reducing direct glare from windows are as follows:
   a. Install an outdoor overhang above the window in question.
   b. Use only those windows that are set at a quite reasonable distance above the floor.
   c. Use light surrounds.
   d. Use shades, louvers, or blinds.
3. *Reflected glare*: Four of the important measures for reducing the reflected glare are as follows:
   a. Use light-diffusing surfaces (e.g., no glossy paper and flat paint).
   b. Position light sources or work areas in such a manner that the reflected light is not directed toward the eyes.
   c. Lower luminaires' luminance level as much as possible.
   d. Provide effective general illumination.

Over the years, professionals working in the area of illumination have made great strides by establishing effective illumination levels for

Table 3.4 Effective illumination levels for conducting five selected activities

No.	Activity description	Illuminance range in lux
1	Visual tasks conducted occasionally (working spaces)	100–200
2	Visual tasks of high contrast or large size (e.g., reading printed material)	200–500
3	Visual tasks of medium contrast or small size (e.g., poorly printed or reproduced material)	500–1,000
4	Visual tasks of very small size or low contrast (e.g., reading handwritten material in hard pencil on poor-quality paper)	1,000–2,000
5	Visual tasks of low contrast and very small size conducted over a prolonged period of time	2,000–5,000

performing various activities. Effective illumination levels for performing some of these activities are presented in Table 3.4 [12,15,21].

3.10 Workload and workload index

Workload is based on the following two assumptions [13,22,23]:

1. The human operator possesses only a limited capacity for processing information within the time available.
2. Tasks are conducted with respect to time.

Overload occurs when the number of events per unit time exceeds the operator's power for selecting and processing them in an effective manner. Past experience over the years clearly indicates that an overhead operator will make errors, as well as omit responses because he or she has too many things to do within the time available.

Although overload occurs quite often in complex tasks with multiple activities, it also can take place with a single activity. For example, a very difficult single task can overload a person's capacities because of the intense mental effort required to perform it.

There are no absolute workload-related measures. In fact, the existing measures are relative and useful only for making comparisons of two or more situations. Examples include workloads under normal and emergency situations or workloads with two totally different controls and displays.

One considered quite useful workload measure or index is defined as follows [13]:

$$I_w = \frac{T_{re}\,(100)}{T_{av}}, \tag{3.42}$$

where I_w is the workload index, T_{re} is the time required, and T_{av} is the time available.

Note that although this index is considered very useful, its three main shortcomings are that the workload inherent to continuous control tasks is not taken into consideration, the workload involved in doing two or more things simultaneously is not considered, and the cognitive activity is not considered [23].

PROBLEMS

1. Describe the bathtub hazard rate curve.
2. Write down the general formulas for the following three functions:
    i.   Failure (or probability) density function
    ii. Reliability function
    iii. Hazard rate function
3. Write down three formulas for obtaining the mean time to failure.
4. Assume that an engineering system is composed of three identical and independent subsystems, and that the constant failure rate of a subsystem is 0.0008 failures/h. All three subsystems must operate normally in order for the engineering system to operate successfully. Calculate the following:
    i.   The engineering system's failure rate
    ii. The engineering system's reliability for a 200-h mission
    iii. The engineering system's mean time to failure
5. What are the special case networks of the k-out-of-n network? Prove Equation 3.34 step by step by using Equation 3.33.
6. What are the typical human behaviors? List at least eight of them.
7. Describe the following human sensory capacities:
    i.   Sight
    ii. Touch
    iii. Noise (hearing)
8. Discuss the effects of vibrations on humans.
9. Describe the following terms:
    i.   Blinding glare
    ii. Disability glare
    iii. Discomfort glare
10. Define the workload index that is considered quite useful. What are its main shortcomings?

References

1. Layman, W. J., Fundamental Consideration in Preparing a Master Plan, *Electrical World*, Vol. 101, 1933, pp. 778–792.
2. Smith, S. A., Service Reliability Measured by Probabilities of Outage, *Electrical World*, Vol. 103, 1934, pp. 371–374.

3. Dhillon, B. S., *Power System Reliability, Safety, and Management*, Ann Arbor Science Publishers, Ann Arbor, MI, 1983.

4. Dhillon, B. S., *Design Reliability: Fundamentals and Applications*, CRC Press, Boca Raton, FL, 1999.

5. Dale Huchingson, R., *New Horizons for Human Factors in Design*, McGraw-Hill, New York, 1981.

6. Dhillon, B. S., *Human Reliability: With Human Factors*, Pergamon Press, New York, 1986.

7. Kapur, K. C., Reliability and Maintainability, in *Handbook of Industrial Engineering*, edited by G. Salvendy, John Wiley and Sons, New York, 1982, pp. 8.5.1–8.5.34.

8. Dhillon, B. S., Life Distributions, *IEEE Transactions on Reliability*, Vol. 30, No. 5, 1981, pp. 457–460.

9. Shooman, M. L., *Probabilistic Reliability: An Engineering Approach*, McGraw-Hill, New York, 1968.

10. Sandler, G. H., *System Reliability Engineering*, Prentice Hall, Englewood Cliffs, NJ, 1963.

11. Lipp, J. P., Topology of Switching Elements Versus Reliability, *Transactions on IRE Reliability and Quality Control*, Vol. 7, 1957, pp. 21–34.

12. Chapanis, A., *Human Factors in Systems Engineering*, John Wiley and Sons, New York, 1996.

13. Dhillon, B. S., *Engineering Usability: Fundamentals, Applications, Human Factors, and Human Error*, American Scientific Publishers, Stevenson Ranch, CA, 2004.

14. Woodson, W. E., *Human Factors Design Handbook*, McGraw-Hill, New York, 1981.

15. McCormick, E. J., and Sanders, M. S., *Human Factors in Engineering and Design*, McGraw-Hill, New York, 1982.

16. *Human Engineering Design Criteria for Military Systems, Equipment, and Facilities*, MIL-STD-1472, Department of Defense, Washington, DC.

17. *Engineering Design Handbook: Maintainability Guide for Design*, AMCP 706-134, Prepared by the United States Army Material Command, 5001 Eisenhower Avenue, Alexandria, VA, 1972.

18. Lederman, S., Heightening Tactile Impression of Surface Texture, in *Active Touch*, edited by G. Gordon, Pergamon Press, New York, 1978, pp. 40–45.

19. *Engineering Design Handbook: Maintainability Engineering Theory and Practice*, AMCP 706-133, Prepared by the United States Army Material Command, 5001 Eisenhower Avenue, Alexandria, VA, 1976.

20. Hultgren, G., and Knave, B., Discomfort Glare and Disturbances from Light Reflections in an Office Landscape with CRT Display Terminals, *Applied Ergonomics*, Vol. 5, No. 1, 1974, pp. 2–8.

21. RQQ Report No. 6, Selection of Illuminance Values for Interior Lighting Design, *Journal of Illuminating Engineering Society*, Vol. 9, No. 3, 1980, pp. 188–190.

22. Lane, D. M., Limited Capacity, Attention Allocation, and Productivity, in *Human Performance and Productivity*, edited by W.C. Howell and E.A. Fleishman, Lawrence Erlbaum Associates, Hillsdale, NJ, 1982, pp. 121–156.

23. Adams, J. A., *Human Factors Engineering*, Macmillan, New York, 1989.

chapter four

Reliability and usability evaluation methods

4.1 Introduction

Over the years, a large amount of published literature in the areas of reliability and usability has appeared in the form of journal articles, conference proceedings articles, technical reports, and books [1–5]. Many of these publications report the development of various types of methods and techniques for performing reliability and usability analysis. Some of these methods and techniques can be utilized quite effectively for performing analysis in both reliability and usability. The others are more confined to a specific area (i.e., reliability or usability).

Three examples of these methods and techniques that can be utilized in the reliability and usability areas are failure modes and effect analysis (FMEA), fault tree analysis (FTA), and the Markov method. FMEA was developed in the early 1950s for analyzing the reliability of engineering systems. Similarly, FTA was developed in the early 1960s for analyzing the safety of rocket launch control systems. Nowadays, both FMEA and FTA are being employed across many diverse areas for analyzing various types of problems.

The Markov method, named after its developer, a Russian mathematician, Andrei A. Markov (1856–1922), is a highly mathematical procedure that is frequently utilized to perform a number of types of reliability and usability analysis in engineering systems.

This chapter presents a number of methods considered useful to perform engineering systems reliability and usability evaluation studies. All of these have been extracted from the published literature in the areas of reliability and usability.

4.2 Failure modes and effect analysis (FMEA)

This is a widely used design tool for analyzing the reliability of engineering systems, and it may simply be described as an approach for analyzing the effects of potential failure modes in the system [1,6]. The history of FMEA goes back to the early 1950s with the development of flight control systems, when the U.S. Navy's Bureau of Aeronautics developed a requirement called *failure analysis* to establish a mechanism for reliability control over the detail design efforts [7]. Eventually, the term was changed to *failure modes and effect analysis* (FMEA).

The seven main steps that are generally followed to perform FMEA are as follows [1,8]:

1. *Step 1*: Define system boundaries and detailed requirements.
2. *Step 2*: List all system components/parts and subsystems.
3. *Step 3*: List each component's/part's failure modes, description, and identification.
4. *Step 4*: Assign a failure occurrence probability/rate to each component/ part failure mode.
5. *Step 5*: List each failure mode effect or effects on subsystems, system, and plant.
6. *Step 6*: Enter appropriate remarks for each failure mode.
7. *Step 7*: Review each critical failure mode and take appropriate actions.

There are many factors that must be explored with care prior to the implementation of FMEA. Some of these factors are as follows [9,10]:

- Examination of each conceivable failure mode by all the involved professionals
- Obtaining the engineer's approval and support
- Making all decisions based on the risk priority number
- Measuring the costs and benefits

Over the years, professionals directly or indirectly involved with reliability analysis have established certain facts and guidelines concerning FMEA. Four of these are shown in Figure 4.1 [8,9].

There are many advantages of performing FMEA. Some of the main ones are as follows [1,8–10]:

- A useful tool for improving communications among design interface personnel
- A systematic approach to categorize and classify hardware failures
- A useful approach for comparing designs and identifying safety concerns
- A useful approach for reducing engineering changes and improving the efficiency of test planning
- A useful approach that starts from the detailed level and works upward
- A helpful tool for safeguarding against repeating the same mistakes in the future

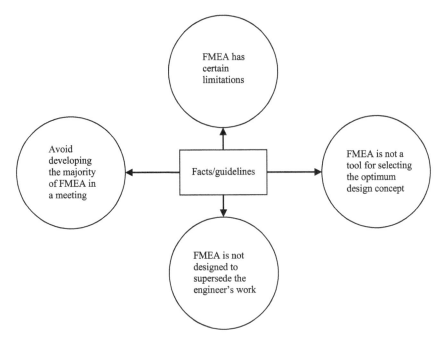

Figure 4.1 FMEA-related facts/guidelines.

- A visibility tool for management that lowers product development time and cost
- A useful tool to understand and improve customer satisfaction

4.3 Fault tree analysis (FTA)

FTA is a widely used method in the industrial sector for evaluating the reliability of engineering systems during their design and development phase, particularly in the area of nuclear power generation. A *fault tree* may simply be described as a logical representation of the relationship of basic events that result in a specified undesirable event, called the *top event*. The fault tree is depicted using a tree structure, with OR, AND, and other logic gates.

This method was developed in the early 1960s at the Bell Telephone Laboratories for conducting analysis of the Minuteman Launch Control System [1,2]. Some of the main objectives of conducting FTA are as follows [1,8]:

- To understand the degree of protection that the design concept provides against the occurrence of failures

- To comprehend the functional relationship of system failures
- To confirm the system's ability to satisfy its imposed safety-related requirements
- To satisfy jurisdictional-related requirements
- To highlight critical areas and cost-effective improvements

There are many prerequisites associated with FTA. Some of the main ones are as follows [1,8]:

- Clearly defined analysis objectives and scope
- Clear identification of all associated assumptions
- A clear definition of what constitutes system or item failure (i.e., the undesirable event)
- A comprehensive review of a system's or item's operational experience
- Thorough comprehension of design, operation, and maintenance aspects of system or item under consideration
- Clearly defined physical bounds and interfaces of a system or item

FTA starts by identifying the top event, which is associated with a system or item under consideration. Fault events that can cause the top event's occurrence are generated and connected by logic operators such as OR and AND. The OR gate provides a true output (i.e., fault) when one or more inputs are true. Similarly, the AND gate provides a true output (i.e., fault) when *all* the inputs are true.

A fault tree's construction proceeds by generating fault events in a successive manner until the fault events need not be developed any further. These fault events are called *primary* or *basic events*. A fault tree relates the top event to the primary/basic fault events. During a fault tree's construction, one question that is successively asked is, "How could this fault event occur?"

Figure 4.2 shows four basic symbols used to construct fault trees. The meanings of the symbols/gates OR and AND, shown in Figure 4.2, have already been discussed. The remaining two symbols (i.e., rectangle and circle) are described next:

- *Rectangle*: It denotes a resultant event that occurs from the combination of fault events through the input of a logic gate such as AND or OR.
- *Circle*: It represents a primary/basic fault event (e.g., the failure of an elementary part/component), and the primary/basic fault-event parameters are failure probability, failure rate, unavailability, and repair rate.

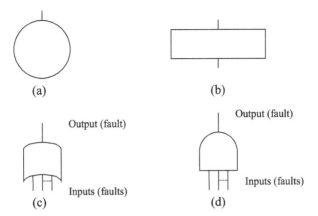

Figure 4.2 Basic fault tree symbols: (a) basic fault event, (b) resultant event, (c) OR gate, (d) AND gate.

EXAMPLE 4.1

Assume that a windowless room contains two light bulbs and one switch. Develop a fault tree for the undesired fault event (i.e., top fault event) "Dark room," if the switch can only fail to close.

In this case, there can be no light in the room (i.e., a dark room) only if both light bulbs burn out, if the switch fails to close, or if there is no incoming electricity. A fault tree for the example, using all four symbols in Figure 4.2, is shown in Figure 4.3. The single capital letters in the fault tree diagram denote corresponding fault events (e.g., T: dark room, Y: both bulbs burned out, and A: power failure).

4.3.1   Fault tree probability evaluation

When the occurrence probabilities of basic/primary fault events are known, the top fault event probability of occurrence can be calculated. This can be done only by first calculating the occurrence probabilities of the output fault events of all the involved intermediate and lower logic gates (e.g., the AND and OR gates).

Thus, the occurrence probability of the AND gate output fault event (say X) is expressed by [1,8]

$$P(X) = \prod_{i=1}^{m} P(x_i),\qquad(4.1)$$

where $P(X)$ is the occurrence probability of the AND gate output fault event X; m is the number of AND gate input independent fault events; and $P(x_i)$ is the probability of occurrence of the AND gate input fault event x_i, for $i = 1, 2, 3, \ldots, m$.

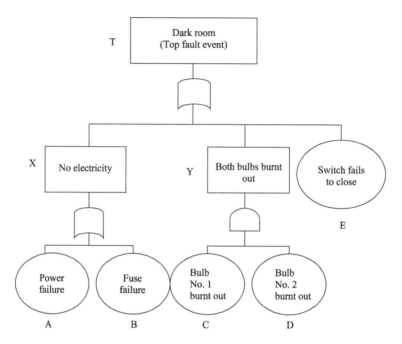

Figure 4.3 A fault tree for the top fault event: Dark room.

Similarly, the occurrence probability of the OR gate output fault event (say Y) is expressed by

$$P(Y) = 1 - \prod_{i=1}^{n} \{1 - P(y_i)\},\qquad(4.2)$$

where $P(Y)$ is the occurrence probability of the OR gate output fault event Y; n is the number of OR gate input independent fault events; and $P(y_i)$ is the probability of occurrence of the OR gate input fault event y_i, for $i = 1, 2, 3, \ldots, n$.

EXAMPLE 4.2

Assume that in Figure 4.3, the occurrence probabilities of independent fault events A, B, C, D, and E are 0.06, 0.05, 0.04, 0.03, and 0.02, respectively. Calculate the probability of occurrence of the top fault event T (Dark room) with the aid of Equations 4.1 and 4.2.

By inserting the specified occurrence probability values of fault events A and B into Equation 4.2, we obtain

$$\begin{aligned}
P(X) &= 1 - [1 - P(A)][1 - P(B)] \\
&= 1 - [1 - 0.06][1 - 0.05] \\
&= 0.107,
\end{aligned}$$

where $P(X)$ is the occurrence probability of fault event X (i.e., no electricity).

Similarly, by inserting the specified occurrence probability values of the fault events C and D into Equation 4.1, we obtain

$$P(Y) = P(C)P(D)$$
$$= (0.04)(0.03)$$
$$= 0.0012,$$

where $P(X)$ is the occurrence probability of fault event Y (i.e., both bulbs burned out).

By inserting these calculated values and the specified data value into Equation 4.2, we obtain

$$P(T) = 1 - [1 - P(X)][1 - P(Y)][1 - P(E)]$$
$$= 1 - [1 - 0.107][1 - 0.0012][1 - 0.02]$$
$$= 0.1259.$$

Thus, the probability of occurrence of the top fault event T (Dark room) is 0.1259.

4.3.2 Advantages and disadvantages of the fault tree analysis

There are many advantages and disadvantages of FTA. Some of its advantages are as follows [1,8]:

- Serves as a very useful graphic aid for system management
- Requires the analyst to understand the system under consideration thoroughly prior to starting the analysis
- Allows the analyst to handle complex systems more easily
- Allows concentration on one specific failure at a time
- Is helpful in providing options for management personnel and others to perform either qualitative or quantitative reliability analysis
- Is a very useful way to highlight failures deductively
- Is a useful way to provide insight into the system behavior.

On the other hand, some of the disadvantages of FTA are as follows [1,8]:

- Results quite difficult to check
- A time-consuming and a costly method
- Considers parts and components in either a functional state or a failed state (i.e., partial-failure states of the parts and components are difficult to handle)

4.4   *Markov method*

The Markov method is frequently used to model engineering systems with constant failure and repair rates. It is subjected to the following assumptions [1,11]:

- The transitional probability from one system state to another in the finite time interval Δt is given by $\alpha \Delta t$, where α is the transition rate (e.g., failure or repair rate) from one system state to another.
- The probability of more than one transition occurrence taking place in the finite time interval Δt is negligible (e.g., $(\alpha \Delta t)(\alpha \Delta t) \to 0$).
- All occurrences are independent of each other.

The application of this method is demonstrated by solving the example presented next.

EXAMPLE 4.3

Assume that an engineering system can be in either an operating or failed state, and its constant failure and repair rates are λ_{es} and μ_{es}, respectively. The system state space diagram is shown in Figure 4.4, and the numerals in circles denote the engineering system states. Develop expressions for the engineering system's time-dependent and steady-state availabilities and unavailabilities, reliability, and mean time to failure by utilizing the Markov method.

Utilizing the Markov method, we write down the following equations for states 0 and 1, shown in Figure 4.4, respectively.

$$P_0(t + \Delta t) = P_0(t)(1 - \lambda_{es}\Delta t) + P_1(t)\mu_{es}\Delta t \qquad (4.3)$$

$$P_1(t + \Delta t) = P_1(t)(1 - \mu_{es}\Delta t) + P_0(t)\lambda_{es}\Delta t, \qquad (4.4)$$

where t is time; $P_0(t + \Delta t)$ is the probability of the engineering system being in operating state 0 at time $(t + \Delta t)$; $P_1(t + \Delta t)$ is the probability

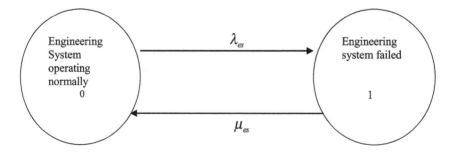

Figure 4.4   Engineering system state space diagram.

of the engineering system being in failed state 1 at time $(t + \Delta t)$; $P_j(t)$ is the probability that the engineering system is in state j at time t, for $j = 0,1$; $\lambda_{es}\Delta t$ is the probability of the engineering system failure in finite time interval Δt; $\mu_{es}\Delta t$ is the probability of the engineering system repair in finite time interval Δt; $(1 - \lambda_{es}\Delta t)$ is the probability of no failure in finite time interval Δt; and $(1 - \mu_{es}\Delta t)$ is the probability of there being no repair in finite time interval Δt.

From Equation 4.3, we obtain

$$P_0(t + \Delta t) = P_0(t) - P_0(t)\lambda_{es}\Delta t + P_1(t)\mu_{es}\Delta t. \tag{4.5}$$

From Equation 4.5, we write

$$\lim_{\Delta t \to 0} \frac{P_0(t + \Delta t) - P_0(t)}{\Delta t} = -P_0(t)\lambda_{es} + P_1(t)\mu_{es}. \tag{4.6}$$

From Equation 4.6, we get

$$\frac{dP_0(t)}{dt} + P_0(t)\lambda_{es} = P_1(t)\mu_{es}. \tag{4.7}$$

Similarly, using Equation 4.4, we obtain

$$\frac{dP_1(t)}{dt} + P_1(t)\mu_{es} = P_0(t)\lambda_{es}. \tag{4.8}$$

At time $t = 0$, $P_0(0) = 1$, and $P_0(0) = 0$.
By solving Equations 4.7 and 4.8, we get [1]

$$P_0(t) = \frac{\mu_{es}}{(\lambda_{es} + \mu_{es})} + \frac{\lambda_{es}}{(\lambda_{es} + \mu_{es})} e^{-(\lambda_{es} + \mu_{es})t} \tag{4.9}$$

$$P_1(t) = \frac{\lambda_{es}}{(\lambda_{es} + \mu_{es})} - \frac{\lambda_{es}}{(\lambda_{es} + \mu_{es})} e^{-(\lambda_{es} + \mu_{es})t}. \tag{4.10}$$

Thus, the engineering system's time-dependent availability and unavailability, respectively, are

$$AV_{es}(t) = P_0(t) = \frac{\mu_{es}}{(\lambda_{es} + \mu_{es})} + \frac{\lambda_{es}}{(\lambda_{es} + \mu_{es})} e^{-(\lambda_{es} + \mu_{es})t} \tag{4.11}$$

$$UA_{es}(t) = P_1(t) = \frac{\lambda_{es}}{(\lambda_{es} + \mu_{es})} - \frac{\lambda_{es}}{(\lambda_{es} + \mu_{es})} e^{-(\lambda_{es} + \mu_{es})t}, \tag{4.12}$$

where $AV_{es}(t)$ is the engineering system's time-dependent availability and $UA_{es}(t)$ is the engineering system's time-dependent unavailability.
By letting time t go to infinity in Equations 4.11 and 4.12, we obtain [1]

$$AV_{es} = \lim_{t \to \infty} AV_{es}(t) = \frac{\mu_{es}}{\lambda_{es} + \mu_{es}} \tag{4.13}$$

and

$$UA_{es} = \lim_{t \to \infty} UA_{es}(t) = \frac{\lambda_{es}}{\lambda_{es} + \mu_{es}}, \tag{4.14}$$

where AV_{es} is the engineering system's steady-state availability and UA_{es} is the engineering system's steady-state unavailability.

For $\mu_{es} = 0$, from Equation 4.9, we obtain

$$R_{es}(t) = P_0(t) = e^{-\lambda_{es}t}, \tag{4.15}$$

where $R_{es}(t)$ is the engineering system reliability at time t.

By integrating Equation 4.15 over the time interval $[0,\infty]$, we obtain the following equation for the engineering system's mean time to failure [1]:

$$
\begin{aligned}
\text{MTTF}_{es} &= \int_0^\infty e^{-\lambda_{es}t} dt \\
&= \frac{1}{\lambda_{es}},
\end{aligned}
\tag{4.16}
$$

where MTTF_{es} is the engineering system's mean time to failure.

Thus, the engineering system's time-dependent and steady-state availabilities and unavailabilities, reliability, and mean time to failure are given by Equations 4.11 through 4.16, respectively.

EXAMPLE 4.4

Assume that the constant failure and repair rates of an engineering system are 0.0002 failures/h and 0.0005 repairs/h, respectively. Calculate the engineering system's steady-state unavailability and unavailability during a 50-h mission.

By inserting the specified data values into Equations 4.14 and 4.12, we obtain

$$UA_{es} = \frac{0.0002}{(0.0002 + 0.0005)} = 0.2857$$

and

$$UA_{es}(50) = \frac{0.0002}{(0.0002 + 0.0005)} - \frac{0.0002}{(0.0002 + 0.0005)} e^{-(0.0002 + 0.0005)(50)}$$

$$= 0.0098.$$

Thus, the engineering system's steady-state unavailability and unavailability during a 50-h mission are 0.2857 and 0.0098, respectively.

4.5 Cognitive walkthroughs

Cognitive walkthroughs are an approach that can be employed for evaluating prototype systems/products. The basic idea behind them is first, to walk through the operation of an interface with all involved individuals, and then to highlight problems within that system. Generally, the method/approach incorporates checklists for use by involved developers to highlight possible problems with an interface. This essentially gives a framework to involved developers for use to check the system from a cognitive perspective. The framework addresses issues such as the following [5,12]:

- Actions to be executed by users at each point in an interaction
- The linking of the interface object components to the actions to be executed by users
- Assumptions with regard to user knowledge

For this method to be used effectively, it is essential to have a clear understanding of the characteristics of all potential users.

Some of the benefits and drawbacks of cognitive walkthroughs are as follows [4]:

- Benefits
  - Are useful for facilitating communication among design personnel, particularly when they are divided into developers and requirement analysts
  - Are useful to comprehend the user's environment
  - Are relatively fast to administer and lead directly to diagnostic and prescriptive information

- Drawbacks
  - Lack of guidance in selecting the proper tasks for evaluation
  - A very high degree of reliance on the judgment of the investigator

Additional information on the method is available in References [4,12–14].

4.6 Cooperative evaluation

The cooperative evaluation method can be utilized during the product/system design's early prototyping stages, or at a later stage when an existing product/system is to be developed further. The method evaluates interfaces based on the use of verbal protocols, with users conducting stated tasks under the designer's observation. More clearly, the tasks are selected by the designer, and he or she observes the user's problems with operating the product/system under study. The method is designed to highlight usability-associated problems with early product/system prototypes, as well as to assist in the design-refining process. The following nine steps are associated with this method [5]:

1. *Step 1*: Recruit appropriate users of the product or system.
2. *Step 2*: Prepare tasks.
3. *Step 3*: Prepare instructions for carrying out an actual usability evaluation.
4. *Step 4*: Ensure the readiness of the prototype for supporting the defined tasks, as well as the availability of recording facilities.
5. *Step 5*: Conduct actual evaluation sessions as appropriate.
6. *Step 6*: Record all relevant data.
7. *Step 7*: Carry out postsession interviews with the users and debrief them as necessary.
8. *Step 8*: Analyze the collected data.
9. *Step 9*: Make necessary recommendations for design improvements.

Some of the main advantages and limitations of this method are as follows [5]:

- Advantages
  - Is useful for improving communication between users and designers
  - Can be utilized by personnel with very little or no training in human factors
  - Is useful for detecting usability-related problems early in the design process

- Limitations
  - Is unsuitable for calculating performance-related data
  - Produces large amounts of data that can be very time consuming to analyze.
  - Is unsuitable for the very early phases of design

4.7 Task analysis

Task analysis is largely used during the product design's specification phase. The method may simply be expressed as the study of what a user of a product is expected to do, with regard to actions and cognitive processes, to achieve a task objective effectively [4]. In general, it may be added that the term *task analysis* refers to a methodology that can be conducted using many specific methods. These methods are basically used for evaluating the interactions between the users and the products and systems.

Task analysis helps to break down the methods for conducting tasks with a product or system under consideration into a series of steps. Consequently, the methods can be utilized to predict whether the performance of tasks in question will be easy or difficult, as well as the degree of effort likely to be needed. The result of basic task analyses provides a list of the physical steps that the user must conduct effectively for completing a specific task. However, it to be noted that complex task analyses also consider the cognitive steps associated with a task.

The number of steps needed for accomplishing a task may be considered as an elementary measure of task complexity. The principle may simply be expressed as "the fewer the steps, the simpler the task." Individuals such as experienced product/system designers, potential product/system users, and domain experts can be valuable informants in conducting task analysis.

Some of the benefits and drawbacks of task analysis are as follows [4,5]:

- Benefits
  - Is useful for highlighting the elements of the design of the product/system that causes the inconsistencies
  - Is useful with regard to prescribing potential solutions to usability-associated problems
  - Requires the investigator to follow a specific procedure because of the standardization of task analysis notations

- Drawbacks
  - Problems with the measure of task complexity (i.e., simply counting the number of steps involved in conducting a task)
  - The assumption of "expert" performance with the system in question

Additional information on this method is available in References [15,16].

4.8   Property checklists

Property checklists basically provide a series of product/system design-related properties as per accepted human factors "wisdom" that will ensure that a product/system under consideration is usable. Generally, the checklists state the usable design's high-level properties, including compatibility, consistency, and good feedback [4]. Subsequently, they list low-level design-associated issues relating to these properties. Some examples of these issues could be the height of characters on a computer screen, labels on products/systems, and specifying the display and control position.

The basic idea behind the property checklists method is to determine if the product/system design properly conforms to the listed properties. If it does not, take appropriate measures to avoid usability-related potential problems. Some of the benefits and drawbacks of the property checklists method are as follows [4]:

- Benefits
  - Can be used throughout the design process and in the evaluation of finished products/systems
  - Is useful for preserving confidentiality because it does not involve user participants
  - Can directly lead to design solutions

- Drawbacks
  - The effectiveness of its actual application depend very much on the judgment of the individuals using it.
  - It is very much subject to the judgment of the individuals compiling the checklist in the first place.

Additional information on this method is available in Reference [17].

4.9   Expert appraisals

The expert appraisals method is based upon the opinions of an expert with regard to product/system usability. The expert possesses extensive experience, particularly with usability-associated issues, with the product/system under consideration. The method may cover very similar issues as the ones covered by the property checklists method, but in greater depth. The main reason for this is that the expert's knowledge allows her or him to focus on really important issues in a particular context. Note that this method sometimes involves more than one expert.

Some of the benefits and drawbacks of the method are as follows [4]:

- Benefits
  - Can lead directly to design solutions.
  - A quite useful approach to provide diagnostic and prescriptive analysis.
  - Does not need any user participants.

- Drawbacks
  - No direct evidence from users that usability-related problems, as highlighted by the expert, will actually lead to problems
  - Total reliance on expert judgment

4.10   Probability tree analysis

Probability tree analysis can be an excellent method for performing usability-related task analysis, by diagrammatically representing human actions and other related events. Diagrammatic task analysis is represented by the probability tree's branches. The tree's branching limbs represent the outcome of each event (i.e., success or failure), and each branch is assigned a probability of occurrence [18].

Some of the benefits of probability tree analysis are flexibility for incorporating (i.e., with some modifications) factors such as emotional stress, interaction stress, and interaction effects; a visibility tool, and simplified mathematical computations.

The method is demonstrated by solving the example presented here.

EXAMPLE 4.5

Assume that a person has to conduct three independent and distinct tasks (a, b, and c) to operate or use an engineering system. Task a is conducted before task b, and task b before task c. Furthermore, each of these tasks can be conducted correctly or incorrectly.

Develop a probability tree and obtain an expression for the probability of not successfully accomplishing the mission (i.e., not operating the engineering system correctly).

In this example, the person first conducts task a correctly or incorrectly, and then proceeds to conduct task b. After conducting task b correctly or incorrectly, the same person proceeds to perform task c.

This whole scenario is depicted by a probability tree, as shown in Figure 4.5.

The symbols used in Figure 4.5 are defined here:

- a denotes the event that task a is conducted correctly.
- $\bar{a}$ denotes the event that task a is conducted incorrectly.
- b denotes the event that task b is conducted correctly.
- $\bar{b}$ denotes the event that task b is conducted incorrectly.
- c denotes the event that task c is conducted correctly.
- $\bar{c}$ denotes the event that task c is conducted incorrectly.

In Figure 4.5, the term abc denotes operating the engineering system successfully (i.e., overall mission success). Thus, the occurrence probability of event abc is [5,19]

$$P(abc) = P_a P_b P_c, \qquad (4.17)$$

where P_a is the probability of conducting task a correctly, P_b is the probability of conducting task b correctly, and P_c is the probability of conducting task c correctly.

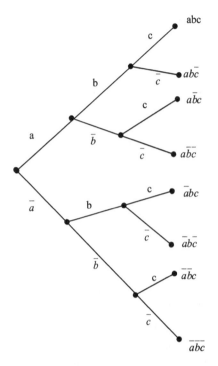

Figure 4.5   A probability tree for conducting tasks a, b, and c.

Similarly, in Figure 4.5, the terms $ab\bar{c}$, $a\bar{b}c$, $a\bar{b}\bar{c}$, $\bar{a}bc$, $\bar{a}b\bar{c}$, $\bar{a}\bar{b}c$, and $\bar{a}\bar{b}\bar{c}$ denote seven distinct possibilities of not operating the engineering system correctly or unsuccessfully. Thus, the probability of not successfully accomplishing the overall mission is

$$P_{ns} = P(ab\bar{c} + a\bar{b}c + a\bar{b}\bar{c} + \bar{a}bc + \bar{a}b\bar{c} + \bar{a}\bar{b}c + \bar{a}\bar{b}\bar{c})$$
$$= P_aP_bP_{\bar{c}} + P_aP_{\bar{b}}P_c + P_aP_{\bar{b}}P_{\bar{c}} + P_{\bar{a}}P_bP_c + P_{\bar{a}}P_bP_{\bar{c}} + P_{\bar{a}}P_{\bar{b}}P_c + P_{\bar{a}}P_{\bar{b}}P_{\bar{c}},$$
$$(4.18)$$

where P_{ns} is the probability of not successfully accomplishing the overall mission (i.e., mission failure); $P_{\bar{a}}$ is the probability of conducting task a incorrectly; $P_{\bar{b}}$ is the probability of conducting task b incorrectly; and $P_{\bar{c}}$ is the probability of conducting task c incorrectly.

Thus, Equation 4.18 is the expression for probability of not successfully accomplishing the mission (i.e., not operating the engineering system correctly).

EXAMPLE 4.6

Assume that in Example 4.5, the probabilities of the person not conducting tasks a, b, and c correctly are 0.3, 0.2, and 0.1, respectively.

Calculate the probability of correctly operating/using the engineering system by this person.

Thus, we have $P_{\bar{a}} = 0.3$, $P_{\bar{b}} = 0.2$, and $P_{\bar{c}} = 0.1$.

Because $P_{\bar{a}} + P_a = 1$, $P_{\bar{b}} + P_b = 1$, and $P_{\bar{c}} + P_c = 1$, we get

$$P_a = 1 - P_{\bar{a}} \qquad (4.19)$$

$$P_b = 1 - P_{\bar{b}} \qquad (4.20)$$

$$P_c = 1 - P_{\bar{c}}. \qquad (4.21)$$

Using Equations 4.19 through 4.21 and the specified data values in Equation 4.17, we obtain

$$\begin{aligned} P(abc) &= (1 - P_{\bar{a}})(1 - P_{\bar{b}})(1 - P_{\bar{c}}) \\ &= (1 - 0.3)(1 - 0.2)(1 - 0.1) \\ &= 0.504. \end{aligned}$$

Thus, the probability of correctly operating or using the engineering system by the person is 0.504.

4.11   Cause-and-effect diagram (CAED)

The cause-and-effect diagram (CAED) method was developed in the 1950s for use in quality control by a Japanese professor, Kaoru Ishikawa, and it can also be used to study usability-related problems. The method is also known as the *Ishikawa diagram* or *fishbone diagram* because of its resemblance to a fish's skeleton, as shown in Figure 4.6. The right side of the diagram (i.e., the "fish head") represents the effect, and the left side shows all the possible causes, which are connected to the central line, called the "fish spine."

The method's main objective is to act as a first step in problem-solving by generating a comprehensive list of expected potential causes. In turn, this can result in the identification of main causes, and thus possible appropriate remedial measures. At a minimum, the application of this method will result in better comprehension of the problem under consideration.

A CAED can be developed by following the four steps presented here:

1. *Step I*: Establish a problem statement and brainstorm to highlight all possible causes.
2. *Step II*: Establish classifications of the main causes by stratifying them into natural groupings and process steps.
3. *Step III*: Develop the diagram by connecting the causes under appropriate process steps and fill in the problem or the effect in the diagram box (i.e., the fish head).

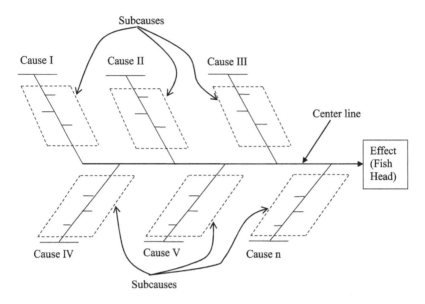

Figure 4.6   A CAED for n number of causes.

4. *Step IV*: Refine the cause classifications by asking questions such as those presented here:
   a.   What caused this?
   b.   What is the main reason for the existence of this condition?

There are many benefits of the CAED method, including the following [5]:

- Very useful for generating ideas
- An effective tool for presenting an orderly arrangement of theories
- Quite useful to highlight root causes
- A quite useful tool in guiding further inquiry

Additional information on this method is available in References [20–22].

EXAMPLE 4.7

Assume that an item or product is being designed by a company for use in an engineering system, and an investigation revealed that it may experience usability problems due to the following five main causes:

1. *Cause I*: Inadequate design and development time.
2. *Cause II*: Poor considerations to human factors during the design process.

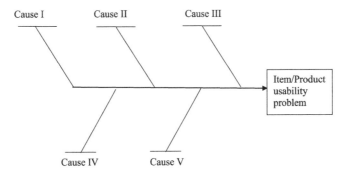

Figure 4.7   A CAED for Example 4.7.

3. *Cause III*: Poor user testing.
4. *Cause IV*: Poor user training.
5. *Cause V*: Poorly written operating instructions.

Construct a CAED.
The CAED for this example is shown in Figure 4.7.

PROBLEMS

1. Describe FMEA.
2. What are the main advantages of performing FMEA?
3. What are the four basic symbols used to construct fault trees? Describe each of these symbols.
4. What are the objectives, advantages, and disadvantages of performing FTA?
5. Prove Equations 4.9 and 4.10 by using Equations 4.7 and 4.8.
6. What are the advantages and disadvantages of cognitive walkthroughs?
7. Describe the following two methods:
    i. Task analysis
    ii. Property checklists
8. Describe the cooperative evaluation method and its advantages and limitations.
9. Compare FTA with probability tree analysis.
10. Describe the CAED and its benefits.

References

1. Dhillon, B. S., *Design Reliability: Fundamentals and Applications*, CRC Press, Boca Raton, FL, 1999.

2. Dhillon, B. S., and Singh, C., *Engineering Reliability: New Techniques and Applications*, John Wiley & Sons, New York, 1981.
3. Jordon, P. W., Thomas, B., and McClelland, I. L., *Issues for Usability Evaluation in Industry: Seminar Discussions*, edited by P. W. Jordon, B. Thomas, B. A. Weerdmeester, and I. L. McClelland, Taylor and Francis Ltd., London, 1996, pp. 237–243.
4. Jordon, P. W., *An Introduction to Usability*, Taylor and Francis Ltd., London, 1998.
5. Dhillon, B. S., *Engineering Usability: Fundamentals, Applications, Human Factors, and Human Error*, American Scientific Publishers, Stevenson Ranch, CA, 2004.
6. Dhillon, B. S., *Systems Reliability, Maintainability, and Management*, Petrocelli Books, New York, 1983.
7. *General Specification for Design, Installation, and Test of Aircraft Flight Control Systems*, MIL-F-18372 (Aer), Bureau of Naval Weapons, Department of the Navy, Washington, DC.
8. Dhillon, B. S., *Transportation Systems Reliability and Safety*, CRC Press, Boca Raton, FL, 2011.
9. Palady, P., *Failure Modes and Effects Analysis*, PT Publications, West Palm Beach, FL, 1995.
10. McDermott, R. E., Mikulak, R. J., and Beauregard, M. R., *The Basics of FMEA*, Quality Resources, New York, 1996.
11. Shooman, M. L., *Probabilistic Reliability: An Engineering Approach*, McGraw-Hill, New York, 1968.
12. Wharton, C., Bradford, J., Jeffries, R., and Franzke, M., Applying Cognitive Walkthroughs to More Complex User Interfaces: Experiences, Issues, and Recommendations, Proceedings of the ACM Conference on Human Factors in Computer Systems, 1992, pp. 381–388.
13. Wharton, C., Rieman, J., Lewis, C., and Polson, P., The Cognitive Walkthrough: A Practitioner's Guide, in *Usability Inspection Methods*, edited by J. Nielsen and R.L. Mack, John Wiley & Sons, New York, 1994, pp. 80–100.
14. Karat, C. M., Campbell, R., and Fiegel, T., Comparison of Empirical Testing and Walkthrough Methods in User Interface Evaluation, Proceedings of the ACM Conference on Human Factors in Computing Systems, 1992, pp. 397–404.
15. Drury, C. G., Task Analysis Methods in Industry, *Applied Ergonomics*, Vol. 14, No. 1, 1983, pp. 19–28.
16. Kirwan, B., and Ainsworth, L. K., editors, *A Guide to Task Analysis*, Taylor and Francis Ltd., London, 1992.
17. Ravden, S. J., and Johnson, G. I., *Evaluating Usability of Human-Interfaces: A Practical Method*, Ellis Horwood Publishers, Chichester, UK, 1989.
18. Swain, A. D., *A Method for Performing a Human Factors Reliability Analysis*, Report No. SCR-685, Sandia Corporation, Albuquerque, NM, August 1963.
19. Dhillon, B. S., *Human Reliability: with Human Factors*, Pergamon Press, New York, 1986.

20. Mears, P., *Quality Improvement Tools and Techniques*, McGraw-Hill, New York, 1995.
21. Kanji, G. K., and Asher, M., *100 Methods for Total Quality Management*, Sage Publications Ltd., London, 1996.
22. Ishikawa, K., *Guide to Quality Control*, Asian Productivity Organization, Tokyo, 1982.

chapter five

Robot system and medical equipment reliability

5.1 Introduction

Nowadays, robots are increasingly being used to perform various types of tasks, including materials handling, arc welding, routing, and spot welding. As robots/robot systems use mechanical, hydraulic, pneumatic, electrical, and electronic parts, their reliability-associated problems are very challenging because of the many sources of failure. A robot/robot system installation is considered successful only if it is safe and reliable. A robot with poor reliability leads to problems such as unsafe conditions, high maintenance-related costs, and inconvenience. Needless to say, robot/robot system reliability has become a very important issue, and over the years, a large number of publications have appeared on the topic, most of which are listed in References 1–4.

Nowadays, billions of dollars are spent to produce various types of medical equipment/devices around the globe [5,6]. The beginning of the medical equipment/device reliability field may be traced to the latter part of the 1960s, when a number of publications that were directly or indirectly concerned with the field appeared [7–11]. These publications covered topics such as "the effect of medical test instrument reliability on patient risks," "safety and reliability in medical electronics," and "instrument-induced errors in the electrocardiogram" [8,10,11]. Needless to say, over the years, a large number of publications on the topic have appeared, most of which are listed in References 12–14.

This chapter presents various important aspects of robot system and medical equipment reliability.

5.2 Robot failure causes and classifications

There are many causes for the occurrence of robot failures. Some the common ones are as follows [2,6]:

- Servo valve–related problems
- Printed circuit board–related problems
- Noise

- Oil pressure valve–related problems
- Encoder-related problems
- Human error

The robot problems/troubles followed the order presented in Table 5.1 [15,16].

Robot failures can be categorized under the four classifications presented here [2,4,17,18]:

1. *Classification I: Systematic hardware faults*—These failures occur because of the existence of unrevealed mechanisms in the design of the robot. Reasons such as peculiar wrist orientations and unusual joint-to-straight-line mode transition can lead the robot not to conduct a specific task or execute specific portions of a program.
2. *Classification II: Random component failures*—These failures occur during the useful life of a robot. They are called *random component failures* because they occur unpredictably. Some of the reasons for the occurrence of such failures are unavoidable failures, undetectable defects, low safety factors, and unexplainable causes.
3. *Classification III: Software failures*—These failures are an important element in the malfunctioning of robots. In robots, software failures and errors can occur in the embedded software or the controlling software and application software. The application of methods such as fault tree analysis (FTA), failure modes and effect analysis (FMEA), and testing can be quite useful for reducing the occurrence of software failures and errors [4,19,20].
4. *Classification IV: Human errors*—These failures are caused by personnel who design, manufacture, operate, and maintain robots. Some of the causes for the occurrence of human errors are poor equipment design, poorly written operating and maintenance procedures, poor

Table 5.1 Robot problems/troubles, in descending order

Number	Problem/trouble
1	Control system problems
2	Incompatibility of jigs and other tools
3	Robot body–related problems
4	Programming and operation errors
5	Welding gun troubles and difficulties with other tooling parts
6	Deterioration, precision deficiency
7	Runaway
8	Miscellaneous

training of operating and maintenance personnel, high temperature in the work area, inadequate lighting in the work area, task complexities, and improper tools [21].

5.3 Factors dictating robot effectiveness and reliabilty measures

Many factors dictate robot effectiveness, including the following [2,4,22]:

- The percentage of time that the robot operates normally
- The percentage of time that the robot is available for service
- The robot's mean time between failures
- The robot's mean time to repair
- The relative performance of the robot under extreme conditions
- The rate of availability of the required spare parts/components
- The quality and availability of personnel required for keeping the robot in an operating state
- The quality and availability of the robot repair equipment and facilities

There are various types of robot reliability-related measures. Four of these measures are described next [2,4,6,20].

5.3.1 Mean time to robot-related problems

The mean time to robot-related problems is the average productive robot time prior to the occurrence of a robot-related problem. It is defined by

$$\text{MTTRP} = \frac{\text{RPT} - \text{DTDRP}}{\text{NRP}}, \tag{5.1}$$

where MTTRP is the mean time to robot-related problems, RPT is the robot production time expressed in hours, DTDRP is the downtime due to robot-related problems expressed in hours, and NRP is the number of robot-related problems.

EXAMPLE 5.1

Assume that at an industrial facility, annual robot production hours and downtime due to robot-related problems are 100,000 h and 1,500 h, respectively. During that period, there were 30 robot-related problems. Calculate the mean time to robot-related problems.

By substituting the specified data values into Equation 5.1, we obtain

$$\text{MTTRP} = \frac{100{,}000 - 1{,}500}{30} = 3{,}283.3 \text{ h}$$

Thus, the mean time to robot-related problems is 3,283.3 h.

5.3.2   *Mean time to robot failure (MTTRF)*

This can be obtained by using any of the following three equations:

$$\text{MTTRF} = \int_0^\infty R_r(t)dt \tag{5.2}$$

$$\text{MTTRF} = \lim_{s \to 0} R_r(s) \tag{5.3}$$

$$\text{MTTRF} = \frac{\text{RPT} - \text{DTDRF}}{\text{NRF}}, \tag{5.4}$$

where MTTRF is the mean time to robot failure, $R_r(t)$ is the robot reliability at time t, $R_r(s)$ is the Laplace transform of the robot reliability function, $R_r(t)$, RPT is the robot production time expressed in hours, DTDRF is the downtime due to robot failures expressed in hours, and NRF is the number of robot failures.

EXAMPLE 5.2

Assume that the constant failure rate, λ_r, of a robot is 0.0004 failures/h, and its reliability is expressed by

$$\begin{aligned} R_r(t) &= e^{-\lambda_r t} \\ &= e^{-(0004)t}, \end{aligned} \tag{5.5}$$

where $R_r(t)$ is the robot reliability at time t.

Calculate MTTRF by using Equations 5.2 and 5.3. Comment on the final result.

By inserting Equation 5.5 into Equation 5.2, we obtain

$$\text{MTTRF} = \int_0^\infty e^{-(0.0004)t} dt$$

$$= \frac{1}{0.0004}$$

$$= 2{,}500 \text{ h.}$$

By taking the Laplace transform of Equation 5.5, we get

$$R_r(s) = \frac{1}{(s + 0.0004)}. \tag{5.6}$$

By inserting Equation 5.6 into Equation 5.3, we obtain

$$\text{MTTRF} = \lim_{s \to 0} \frac{1}{(s + 0.0004)}$$

$$= \frac{1}{0.0004}$$

$$= 2{,}500 \text{ h.}$$

In both cases, the final result (i.e., MTTRF = 2,500 h) is the same. It proves that Equations 5.2 and 5.3 yield exactly the same result.

EXAMPLE 5.3

Assume that the annual production hours of a robot and its annual downtime due to failures are 50,000 h and 400 h, respectively. During that period, the robot failed 10 times. Calculate the MTTRF.

By inserting the given data values into Equation 5.4, we get

$$\text{MTTRF} = \frac{50{,}000 - 400}{10}$$

$$= 4{,}960 \text{ h}$$

Thus, the MTTRF is 4,960 h.

5.3.3   Robot reliability

Robot reliability may simply be described as the probability that a robot will carry out its specified function satisfactorily for the stated time period when used as per designed conditions. The general formula for obtaining time-dependent robot reliability is expressed as follows [2,4,20]:

$$R_r(t) = \exp\left[-\int_0^t \lambda_r(t)dt\right],$$ (5.7)

where $R_r(t)$ is the robot reliability at time t and $\lambda_r(t)$ is the robot hazard rate or time-dependent failure rate.

Equation 5.7 is the general expression for obtaining robot reliability. It can be used to obtain the reliability function of a robot for any failure times probability distribution (e.g., gamma, Weibull, or exponential).

EXAMPLE 5.4

Assume that the hazard rate of a robot is defined by

$$\lambda_r(t) = \frac{\beta t^{\beta-1}}{\theta\theta^{\beta-1}},$$ (5.8)

where $\lambda_r(t)$ is the hazard rate or time-dependent failure rate of the robot when its times to failure follow the Weibull distribution; θ is the scale parameter; β is the shape parameter; and t is time.

Obtain an expression for the robot reliability in this instance and then use it to calculate reliability when $t = 400$ h, $\beta = 1$, $\theta = 800$ h.

By inserting Equation 5.8 into Equation 5.7, we obtain

$$R_r(t) = \exp\left[-\int_0^t \frac{\beta t^{\beta-1}}{\theta\theta^{\beta-1}}dt\right]$$ (5.9)

$$= e^{-(t/\theta)^\beta}.$$

By inserting the specified data values into Equation 5.9, we obtain

$$R_r(400) = e^{-(400/800)}$$

$$= 0.6065.$$

Thus, Equation 5.9 is the expression for robot reliability, and the robot reliability for the specified mission period of 400 h is 0.6065.

5.3.4   Robot hazard rate

The robot hazard rate is defined as follows [2,4,20]:

$$\lambda_r(t) = -\frac{1}{R_r(t)}\frac{dR_r(t)}{dt},$$ (5.10)

where $\lambda_r(t)$ is the robot hazard rate and $R_r(t)$ is the robot reliability at time t.

Equation 5.10 can be used to obtain the robot hazard rate when robot times to failure follow any time-continuous probability distribution (e.g., exponential, Weibull, Rayleigh, etc.).

EXAMPLE 5.5

Assume that the reliability of a robot is defined by

$$R_r(t) = e^{-\lambda_r t}, \tag{5.11}$$

where $R_r(t)$ is the robot reliability at time t and λ_r is the robot constant failure rate.

Obtain an expression for the robot hazard rate by using Equations 5.10 and 5.11.

By inserting Equation 5.11 into Equation 5.10, we obtain

$$
\begin{aligned}
\lambda_r(t) &= -\frac{1}{e^{-\lambda_r t}} \frac{de^{-\lambda_r t}}{dt} \\
&= -\frac{1}{e^{-\lambda_r t}} [-\lambda_r e^{-\lambda t}] \\
&= \lambda_r.
\end{aligned}
\tag{5.12}
$$

Thus, Equation 5.12 is the expression for the robot hazard rate, which is constant (i.e., it does not depend on time).

5.4 Robot reliability analysis methods and models

In the field of reliability engineering, there are many methods used for performing various types of reliability analysis [20]. Some of these methods can be used quite effectively to perform robot reliability–related studies. Four of these methods are as follows:

1. *Method I: Parts count method*—This method is generally used during a bid proposal and early design phases for estimating the failure rate of the system or equipment. The method requires information on items such as the use environment of the system or equipment, quantities and types of generic parts, and quality levels of parts.

   Additional information on the method is available in References [20,23].
2. *Method II: FMEA*—This method was developed in the early 1950s by the U.S. Department of Defense and is considered a very effective tool for performing analysis of each failure mode in the system/equipment to determine the effects of such a failure mode on

the entire system/equipment [24]. The method is composed of six steps:

a. define system/equipment and its associated requirements
b. develop appropriate ground rules
c. describe the system/equipment and all its related functional blocks
d. highlight all possible failure modes and their effects
e. develop critical items' list
f. document the analysis [24,25]

Additional information on this method is available in Chapter 4 and in References 20,24, and 25.

3. *Method III: FTA*—This method is widely used to evaluate the reliability of engineering systems during their design and development phase. It was developed at the Bell Telephone Laboratories in the early 1960s. A *fault tree* may be described as a logical representation of the relationship of basic/primary fault events that lead to the occurrence of a specified undesirable event called the *top event*.

Additional information on this method is available in Chapter 4 and in References 20,26, and 27.

4. *Method IV: Markov method*—This method can be utilized in more cases than any other reliability evaluation method and is utilized to model systems with constant failure and repair rates. The method is described in detail in Chapter 4 and in References 20,28.

Over the years, many mathematical models have been developed for performing robot reliability–related studies. Although the effectiveness of these models can vary significantly from one application to another, some are being used quite successfully for performing robot reliability–related studies in the industrial sector.

Many of these models were developed by utilizing the Markov method. Additional information on these models is available in Chapter 13 and in References 2,6, and 29–31.

5.5   Electric and hydraulic robots' reliability analysis

As both electric and hydraulic robots are used in the industrial sector, this scenario is important to explore. This section presents reliability analysis of two typical electric and hydraulic robots by utilizing the block diagram method [2,4,17,18]. Generally, for the purpose of design evaluation, it is assumed that all parts of both electric and hydraulic robots act in series (i.e., if any part fails, the robot fails).

Reliability analyses of both electric and hydraulic robots are presented in the next subsections.

5.5.1   Reliability analysis of an electric robot

For the purposes of this discussion, this electric robot is one that conducts a supposedly normal industrial task, while its programming and maintenance are conducted by humans. The robot is subject to the following seven assumptions/factors [2,4,17]:

1. A direct current (DC) motor actuates each joint.
2. The motor shaft rotation is transmitted to the robot's appropriate limb through a transmission unit.
3. The transducer sends all appropriate signals to the joint controller.
4. A supervising controller/computer directs all joints.
5. An interface bus allows interaction between the joint control processors and the supervisory controller.
6. Each joint is coupled with a feedback transducer (encoder).
7. A microprocessor control card controls each joint.

The block diagram shown in Figure 5.1 represents the electric robot under consideration here with regard to reliability.

Figure 5.1 shows that the electric robot is comprised of two hypothetical subsystems (i.e., subsystem1 and subsystem 2) in series. In turn, as shown in Figures 5.2a and b, subsystem 1 is comprised of two hypothetical elements (i.e., element M and element N) in series, and subsystem 2 is comprised of five parts (i.e., interface, end effector, supervisory controller/computer, joint control, and drive transmission) in series. Furthermore, as shown in Figure 5.2c, element M is comprised of two hypothetical subelements (i.e., subelement C and subelement D) in series.

Consulting Figure 5.1, we get the following expression for the probability of nonoccurrence of the undesirable electric robot movement (reliability):

$$R_{er} = R_{s1}R_{s2},\qquad(5.13)$$

where R_{er} is the probability of nonoccurrence (reliability) of the undesirable electric robot movement, R_{s1} is the reliability of the independent subsystem 1, and R_{s2} is the reliability of the independent subsystem 2.

Figure 5.1   Block diagram for estimating the nonoccurrence probability (reliability) of the electric robot's undesirable movement.

For independent elements, the reliability of subsystem 1 in Figure 5.2a is

$$R_{s1} = R_M R_N, \tag{5.14}$$

where R_M is the reliability of element M and R_N is the reliability of element N.

For independent hypothetical subelements, the reliability of element M in Figure 5.2c is

$$R_M = R_C R_D, \tag{5.15}$$

where R_C is the maintenance person's reliability with regard to causing robot movement (subelement C) and R_D is the maintenance operator's reliability with regard to causing robot movement (subelement D).

Similarly, for independent parts, the reliability of subsystem 2 in Figure 5.2b is

$$R_{s2} = R_1 R_2 R_3 R_4 R_5, \tag{5.16}$$

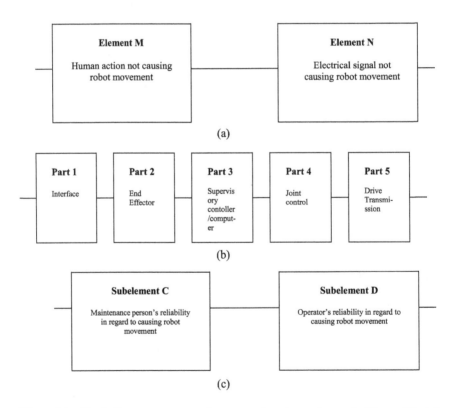

Figure 5.2 Block diagram representing two subsystems shown in Figure 5.1. (a) Subsystem 1, (b) subsystem 2, and (c) element M of subsystem 1.

where R_1 is the reliability of part 1 (i.e., interface); R_2 is the reliability of part 2 (i.e., end effector); R_3 is the reliability of part 3 (i.e., supervisory controller/computer); R_4 is the reliability of part 4 (i.e., joint control); and R_5 is the reliability of part 5 (i.e., drive transmission).

EXAMPLE 5.6

Assume that the following reliability data values are specified for this type of electric robot:

$$R_N = 0.85, \ R_C = 0.91, \ R_D = 0.94, \ R_1 = 0.98, \ R_2 = 0.96,$$
$$R_3 = 0.95, R_4 = 0.94, \ \text{and} \ R_5 = 0.92$$

Calculate the probability of nonoccurrence (reliability) of the undesirable electric robot movement.

By substituting the given data values into Equations 5.15 and 5.16, we obtain

$$R_M = (0.91)(0.94) = 0.8554$$

and

$$R_{s2} = (0.98)(0.96)(0.95)(0.94)(0.92) = 0.7729.$$

By substituting these calculated values for R_M and the specified data value for R_N into Equation 5.14, we obtain

$$R_{s1} = (0.8554)(0.85) = 0.7271.$$

By inserting these calculated values into Equation 5.13, we get

$$R_{er} = (0.7271)(0.7729) = 0.5619.$$

Thus, the probability of nonoccurrence (reliability) of the undesirable electric robot movement is 0.5619.

5.5.2   Reliability analysis of a hydraulic robot

The hydraulic robot considered in this section contains five joints. In turn, each joint is controlled and driven by a hydraulic servomechanism. The robot is subject to the following seven assumptions [2,4,18]:

1. An operator uses a teach pendant to control the arm motion in teach mode.

2. Under high flow demand, an accumulator assists the pump to supply additional hydraulic fluid.
3. Each hydraulic actuator's motion is controlled by a servo valve. This motion is transmitted directly or indirectly (i.e., through chains, rods, gears, etc.) to the robot's specific limb and, in turn, each limb is coupled to a position transducer.
4. A conventional motor and pump assembly generates pressure.
5. An unloading valve is utilized to keep pressure under the maximum limit.
6. The position transducer provides the joint angle codes and, in turn, each code's scanning is carried out by a multiplexer.
7. Hydraulic fluid is pumped from the reservoir.

The block diagram shown in Figure 5.3 represents the hydraulic robot under consideration with regard to reliability. Figure 5.3 shows that the hydraulic robot is comprised of four subsystems [i.e., subsystem (1)—hydraulic pressure supply subsystem, subsystem (2)—drive subsystem, subsystem (3)—gripper subsystem, and subsystem (4)—electronic and control subsystem] in series. As shown in Figures 5.4a–c, the hydraulic pressure supply subsystem is comprised of two parts (parts 1 and 2) in series, the drive subsystem is comprised of five parts (joints 1, 2, 3, 4, and 5) in series, and the gripper subsystem is comprised of two parts (part 1 and part 2) in series, respectively.

Consulting Figure 5.3, we get the following expression for the probability of the nonoccurrence of a hydraulic robot event (i.e., undesirable hydraulic arm movement that causes damage to the robot/other equipment and possible harm to humans):

$$R_{hr} = R_{s1} R_{s2} R_{s3} R_{s4}, \tag{5.17}$$

where R_{hr} is the hydraulic robot reliability or the probability of the nonoccurrence of the hydraulic robot event (i.e., undesirable robot arm movement causing damage to the robot/other equipment and possible

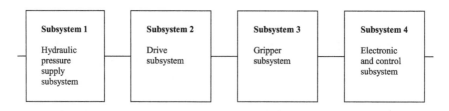

Figure 5.3 Block diagram of the hydraulic robot under consideration.

harm to humans); R_{s1} is the reliability of the independent hydraulic pressure supply subsystem (i.e., subsystem 1); R_{s2} is the reliability of the independent drive subsystem (i.e., subsystem 2); R_{s3} is the reliability of the independent gripper subsystem (i.e., subsystem 3); and R_{s4} is the reliability of the independent electronic and control subsystem (i.e., subsystem 4).

For independent parts, the reliabilities R_{s1}, R_{s2}, and R_{s3} of the hydraulic pressure supply subsystem, drive subsystem, and gripper subsystem using Figures 5.4a–c, respectively, are

$$R_{s1} = R_p R_{hc} \tag{5.18}$$

$$R_{s2} = R_{j1} R_{j2} R_{j3} R_{j4} R_{j5} \tag{5.19}$$

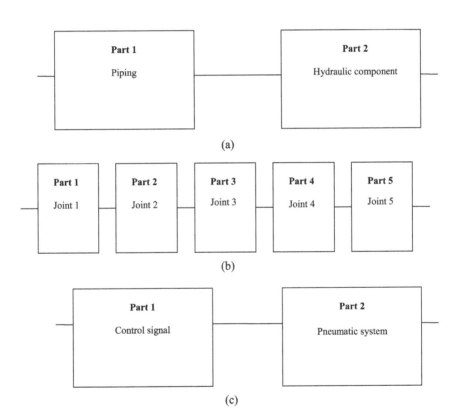

Figure 5.4 Block diagram representing three subsystems shown in Figure 5.3. (a) Hydraulic pressure supply subsystem, (b) drive subsystem, and (c) gripper subsystem.

and

$$R_{s3} = R_{cs}R_{ps}, \tag{5.20}$$

where R_p is the reliability of the piping, R_{hc} is the reliability of the hydraulic component, R_{j1} is the reliability of joint 1, R_{j2} is the reliability of joint 2, R_{j3} is the reliability of joint 3, R_{j4} is the reliability of joint 4, R_{j5} is the reliability of joint 5, R_{cs} is the reliability of the control signal, and R_{ps} is the reliability of the pneumatic system.

For the constant failure rates of independent subsystems shown in Figure 5.3, and, in turn, of their corresponding independent parts shown in Figure 5.4, using Equations 5.17 through 5.20, we get

$$\begin{aligned}
R_{hr}(t) &= e^{-\lambda_{s1}t}e^{-\lambda_{s2}t}e^{-\lambda_{s3}t}e^{-\lambda_{s4}} \\
&= e^{-\lambda_p t}e^{-\lambda_{hc}t}e^{-\lambda_{j1}t}e^{-\lambda_{j2}t}e^{-\lambda_{j3}t}e^{-\lambda_{j4}t}e^{-\lambda_{j5}t}e^{-\lambda_{cs}t}e^{-\lambda_{ps}t}e^{-\lambda_{s4}t} \\
&= e^{-\left(\lambda_p + \lambda_{hc} + \sum_{i=1}^{5}\lambda_{ji} + \lambda_{cs} + \lambda_{ps} + \lambda_{s4}\right)t},
\end{aligned} \tag{5.21}$$

where $R_{hr}(t)$ is the hydraulic robot reliability or the probability of the nonoccurrence of the hydraulic robot event (i.e., undesirable robot arm movement that causes damage to the robot/other equipment and possible harm to humans) at time t; λ_{s1} is the constant failure rate of subsystem 1 (i.e., hydraulic pressure supply subsystem); λ_{s2} is the constant failure rate of subsystem 2 (i.e., drive subsystem); λ_{s3} is the constant failure rate of subsystem 3 (i.e., gripper subsystem); λ_{s4} is the constant failure rate of subsystem 4 (i.e., electronic and control subsystem); λ_p is the constant failure rate of the piping; λ_{hc} is the constant failure rate of the hydraulic component; λ_{j1} is the constant failure rate of joint 1; λ_{j2} is the constant failure rate of joint 2; λ_{j3} is the constant failure rate of joint 3; λ_{j4} is the constant failure rate of joint 4; λ_{j5} is the constant failure rate of joint 5; λ_{cs} is the constant failure rate of the control signal; and λ_{ps} is the constant failure rate of the pneumatic system.

By integrating Equation 5.21 over the time interval $[0, \infty]$, we obtain

$$\begin{aligned}
\text{MTTO}_{hre} &= \int_0^\infty e^{-\left(\lambda_p + \lambda_{hc} + \sum_{i=1}^{5}\lambda_{ji} + \lambda_{cs} + \lambda_{ps} + \lambda_{s4}\right)t}\,dt \\
&= \frac{1}{\lambda_p + \lambda_{hc} + \sum_{i=1}^{5}\lambda_{ji} + \lambda_{cs} + \lambda_{ps} + \lambda_{s4}},
\end{aligned} \tag{5.22}$$

where MTTO_{hre} is the mean time to the occurrence of the hydraulic robot event (undesirable arm movement that causes damage to the robot/other equipment and possible harm to humans).

EXAMPLE 5.7

Assume that the given constant failure rate values for this type of hydraulic robot are $\lambda_p = 0.0008$ failures/h, $\lambda_{hc} = 0.0007$ failures/h, $\lambda_{ji} = \lambda_{j2} = \lambda_{j3} = \lambda_{j4} = \lambda_{j5} = 0.0006$ failures/h, $\lambda_{cs} = 0.0005$ failures/h, $\lambda_{ps} = 0.0004$ failures/h, and $\lambda_{s4} = 0.0003$ failures/h.

Using Equation 5.22, calculate the mean time to the occurrence of the hydraulic robot event (undesirable arm movement that causes damage to the robot/other equipment and possible harm to humans).

By inserting the given data values into Equation 5.22, we get

$$\text{MTTO}_{hre} = \frac{1}{0.0008 + 0.0007 + 5(0.0006) + 0.0005 + 0.0004 + 0.00003)}$$

$$= 175.43 \text{ h}.$$

Thus, the mean time to the occurrence of the hydraulic robot event (undesirable robot arm movement that causes damage to the robot/other equipment and possible harm to humans) is 175.43 h.

5.6   *Facts and figures related to medical equipment reliability*

Some of the facts and figures that are directly or indirectly concerned with medical equipment/device reliability are as follows [6]:

- In 1997, there were 10,420 registered medical device manufacturers in the United States [32].
- Due to faulty instrumentation, approximately 1,200 deaths per year occur in the United States alone [33,34].
- A study reported that more than 50% of all technical medical equipment–related problems were due to operator error [35].
- In 1990, a study conducted by the U.S. Food and Drug Administration (FDA) revealed that approximately 44% of all quality-associated problems that resulted in the voluntary recall of medical devices for the period October 1983–September 1989 were the result of errors or deficiencies that could have prevented through proper design controls [36].
- In 1969, a special committee of the U.S. Department of Health, Education, and Welfare reported that over a 10-year period, approximately 10,000 injuries were associated with medical devices/equipment, with 731 resulting in deaths [37,38].
- The Emergency Care Research Institute (ECRI) tested a sample of 15,000 products used in hospitals and discovered that approximately 4%–6% of these products were sufficiently dangerous to warrant immediate correction [35].

5.7   Medical equipment reliability improvement procedures and methods

There are many procedures and methods used for improving medical equipment reliability. Some of these are described in the next subsections.

5.7.1   General approach

The general approach is a 13-step procedure developed by Bio-Optronics in order to produce reliable and safe medical devices [39]. These steps are as follows [6,39]:

1. *Step 1*: Analyze existing medical problems.
2. *Step 2*: Develop a product concept for determining a possible solution to a specific medical-related problem.
3. *Step 3*: Evaluate all possible environments in which the medical device/equipment will function.
4. *Step 4*: Evaluate all possible personnel expected to operate the device/equipment under consideration.
5. *Step 5*: Construct a prototype.
6. *Step 6*: Test the prototype in laboratory environments.
7. *Step 7*: Test the prototype in actual field use environments.
8. *Step 8*: Make appropriate changes to the product (device/equipment) design to satisfy the field requirements.
9. *Step 9*: Carry out laboratory and field testing on the modified version of the device/equipment.
10. *Step 10*: Build pilot units for performing all necessary tests.
11. *Step 11*: Ask impartial experts to test pilot units in actual field use environments.
12. *Step 12*: Release the product (device/equipment) design for production.
13. *Step 13*: Study the device/equipment field performance and support with appropriate device/equipment maintenance.

5.7.2   Parts count method

The parts count method was briefly discussed in Section 5.4 earlier in this chapter, and additional information is available in References 20,23.

5.7.3   Failure modes and effect analysis (FMEA)

FMEA was briefly discussed in Section 5.4, and additional information on the method is available in Chapter 4 and in References 20,24, and 25.

5.7.4   Fault tree analysis (FTA)

FTA was briefly discussed in Section 5.4, and additional information on the method is available in Chapter 4 and in References 20,26, and 27.

5.7.5   Markov method

The Markov method was briefly discussed in Section 5.4, and additional information on the method is available in Chapter 4 and in References 20,28.

5.8   Human error in medical equipment

Human errors are universal; they are committed each day around the globe. Past experiences over the years indicate that although most of these errors are trivial, some can be quite serious (even fatal). In the area of health care, for instance, one study revealed that in a typical year, about 100,000 Americans die due to human error [40]. Nonetheless, some facts and figures about human errors related, directly or indirectly, to medical equipment and devices are presented here:

- More than 50% of all technical medical equipment/device problems are due to operator error [35].
- The Center for Devices and Radiological Health (CDRH) of the FDA reported that human error accounts for about 60% of all equipment/device-related deaths or injuries in the United States [41].
- A patient was seriously injured by overinfusion because the attending nurse wrongly read the number 7 as 1 [42].
- Human error, directly or indirectly, is responsible for up to 90% of accidents, both generally and in medical devices/equipment [43,44].
- A fatal radiation overdose accident involving the Therac radiation therapy device/equipment was the result of human error [45].

5.8.1   Important medical equipment/device operator-related errors

There are various types of operator-related errors that occur during medical equipment/device operation/maintenance. Some of the important ones are presented here [6,46]:

- Incorrect decision-making and actions in critical moments
- Incorrect selection of devices in regard to the clinical requirements and objectives
- Departure from following stated instructions and procedures
- Misassembly

- Mistakes in setting device/equipment parameters
- Over-reliance on automatic equipment/device features
- Wrong interpretation of or failure to recognize critical device/equipment outputs
- Untimely/inadvertent activation of controls

5.8.2   *Medical devices/equipment with high incidence of human error*

Each day, human errors in using medical devices/equipment in the United States cause at least three deaths or serious injuries [47]. Over the years, many studies have been carried out to identify and highlight medical devices/equipment with a high occurrence of human error. Consequently, the most error-prone medical devices/equipment were highlighted. These devices/equipment, in the order of least error-prone to most error-prone are as follows [47]:

- Least error-prone
  - Contact lens cleaning and disinfecting solutions
  - Continuous ventilator (respirator)
  - External low-energy defibrillator
  - Transluminal coronary angioplasty catheter
  - Catheter guide wire
  - Catheter introducer
  - Peritoneal dialysate delivery system
  - Implantable pacemaker
  - Mechanical/hydraulic impotence device
  - Nonpowered suction apparatus
  - Electrosurgical cutting and coagulation device
  - Urological catheter
  - Infusion pump
  - Intravascular catheter
  - Implantable spinal cord simulator
  - Permanent pacemaker electrode
  - Administration kit for peritoneal dialysis
  - Orthodontic bracket aligner
  - Balloon catheter
- Most error-prone
  - Glucose meter

5.9   *Medical equipment maintenace–related indices*

Medical equipment maintenance may simply be described as all actions required for retaining medical equipment in, or restoring to, a stated condition. Just as in the case of general maintenance activity, there are many indices that can be utilized for measuring the effectiveness of equipment's maintenance activity.

Three of these indices are described next [48].

5.9.1 Index I

Index I measures how frequently the customer has to request service for medical equipment and is expressed by

$$\alpha_c = \frac{n_{rr}}{m},\qquad(5.23)$$

where α_c is the number of repair requests completed per medical equipment, n_{rr} is the total number of repair requests, and m is the number of pieces of medical equipment.

According to one study, the value of α_c ranged from 0.3 to 2 [49].

5.9.2 Index II

Index II measures how much time elapses from a customer request until the failed medical equipment is repaired and put back into full service. The index is defined by

$$\alpha_{at} = \frac{T_t}{n},\qquad(5.24)$$

where α_{at} is the average turnaround time per repair, T_t is the total turnaround time, and n is the total number of work orders/repairs.

The value of α_{at} ranged from 35.4 to 135 h [49].

5.9.3 Index III

Index III is a cost ratio and is defined by

$$\alpha_{cr} = \frac{\text{MEAC}}{\text{MESC}},\qquad(5.25)$$

where α_{cr} is the cost ratio, MEAC is the medical equipment acquisition cost, and MESC is the medical equipment service cost.

Note that this cost includes all labor, parts, and material costs and scheduled and unscheduled service, including in-house, vendor, prepaid contracts, and maintenance insurance.

For various classifications of medical equipment, a range of values for this index are available in Reference 49.

5.10   Guidelines for reliability and other professionals for improving medical equipment reliability

There is a larger number of professionals involved in the manufacture and use of various types of medical equipment/devices. Reliability analysts and engineers are among these professionals. Nonetheless, some useful guidelines for reliability and other professional for improving medical equipment reliability are presented next [6,20,50].

- Guidelines for reliability professionals include the following:
  - Focus on critical failures, as not all equipment/device failures are equally important.
  - Use methods such as FMEA, qualitative FTA, design review, and parts review for obtaining immediate results.
  - Keep in mind that manufacturers are responsible for reliability during the equipment/device design and manufacturing phase; and during its operational phase, it is basically the users' responsibility.
  - Focus on cost-effectiveness, and always keep in mind that some reliability improvement-related decisions need very small or no additional expenditure.
  - Always aim to use simple reliability methods/approaches as much as possible instead of sophisticated methods/approaches used in the aerospace industrial sector.
- Guidelines for other professionals include the following:
  - Compare human body and medical equipment/device failures. Both of them need proper measures from reliability professionals and doctors for enhancing equipment/device reliability and extending human life, respectively.
  - Recognize that failures are the cause of poor medical equipment/ device reliability, and positive thinking and measures can be very useful for improving equipment/device reliability.
  - Keep in mind that the application of reliability principles have successfully improved the reliability of systems used in the aerospace area, and their applications to medical equipment/devices can generate similar dividends.
  - To achieve success with regard to equipment/device reliability, both manufacturers and users must accept their share of associated responsibilities.
  - Remember that the cost of failures in a business organization is probably the largest single expense. Such failures could be associated with items such as equipment, business systems, and people; and a reduction in these failures can decrease the cost of business quite significantly.

5.11   Sources and organizations for obtaining medical eqipment/device failure-related data

There are many sources and organizations from which failure data directly or indirectly concerned with medical equipment/device can be obtained. These sources and organizations are presented next.

- Sources
  - *Medical Device Reporting System (MDRS)*: This system was developed by Center for Devices and Radiological Health [51].
  - *Hospital Equipment Control System (HECS)*: This system was developed by ECRI in 1985 [52].
  - *NUREG/CR-1278: Handbook of Human Reliability Analysis with Emphasis on Nuclear Power Plant Applications*: This standard was published by the U.S. Nuclear Regulatory Commission, Washington, DC.
  - *Universal Medical Device Registration and Regulatory Management System (UMDRMS)*: This system was also developed by ECRI [52].
  - *MIL-HDBK-217: Reliability Prediction of Electronic Equipment*: This standard was published by the Department of Defense, Washington, DC.
- Organizations
  - Emergency Care Research Institute (ECRI), 5200 Butler Parkway, Plymouth Meeting, PA 19462
  - Center for Devices and Radiological Health (CDRH), U.S. Food and Drug Administration (FDA), 1390 Piccard Drive, Rockville, MD 20850
  - Parts Reliability Information Center (PRINCE), Reliability Office, George C. Marshall Space Flight Center, National Aeronautics and Space Administration (NASA), Huntsville, AL 35812
  - National Technical Information Service, 5285 Port Royal Road, Springfield, VA 22161
  - Government Industry Data Exchange Program (GIDEP), GIDEP Operations Center, Fleet Missile Systems, Analysis, and Evaluation Group, Department of Navy, Corona, CA 91720
  - Reliability Analysis Center (RAC), Rome Air Development Center (RADC), Griffis Air Force Base, Department of Defense, Rome, NY 13441

PROBLEMS

1. Discuss classifications of robot failures and their common causes.
2. What are the dictating factors of robot effectiveness?

3. Write down three formulas that can be used to calculate MTTRF.
4. Discuss at least three methods that can be used to conduct robot reliability analysis.
5. Compare an electric robot with a hydraulic robot with respect to reliability.
6. List at least six facts and figures concerned, directly or indirectly, with medical equipment/device reliability.
7. List at least 10 medical devices/equipment with high incidence of human error.
8. Discuss at least eight useful guidelines for reliability and other professionals for improving medical equipment reliability.
9. List at least three good sources and two good organizations for obtaining data related to medical equipment/device reliability.
10. Discuss at least four important operator-related errors that occur during operation and maintenance of medical equipment/ devices.

References

1. Dhillon, B. S., On Robot Reliability and Safety: Bibliography, *Microelectronics and Reliability*, Vol. 27, 1987, pp. 105–118.
2. Dhillon, B. S., *Robot Reliability and Safety*, Springer-Verlag, New York, 1991.
3. Dhillon, B. S., Fashandi, A. R. M. and Liu, K. L., Robot System Reliability: A Review, *Journal of Quality in Maintenance Engineering*, Vol. 8, No. 3, 2002, pp. 170–212.
4. Dhillon, B. S., *Robot System Reliability and Safety: A Modern Approach*, CRC Press, Boca Raton, FL, 2015.
5. Murray, K., Canada's Medical Device Industry Faces Cost Pressures, Regulatory Reform, *Medical Device and Diagnostics Magazine*, Vol. 19, No. 8, 1997, pp. 30–39.
6. Dhillon, B. S., *Applied Reliability and Quality: Fundamentals, Methods, and Procedures*, Springer-Verlag, London, 2007.
7. Johnson, J. P., Reliability of ECG Instrumentation in a Hospital, *Proceedings of the Annual Symposium on Reliability*, 1967, pp. 314–318.
8. Meyer, J. L., Some Instrument-Induced Errors in the Electocardiogram, *Journal of the American Medical Association*, Vol. 201, 1967, pp. 351–358.
9. Gechman, R., Tiny Flaws in Medical Design Can Kill, *Hospital Topics*, Vol. 46, 1968, pp. 23–24.
10. Taylor, E. F., The Effect of Medical Test Instrument Reliability on Patient Risks, *Proceedings of the Annual Symposium on Reliability*, 1969, pp. 328–330.
11. Crump, J. F., Safety and Reliability in Medical Electronics, *Proceedings of the Annual Symposium on Reliability*, 1969, pp. 320–330.
12. Dhillon, B. S. and McCrea, J. L., Bibliography of Literature on Medical Equipment Reliability, *Microelectronics and Reliability*, Vol. 20, 1980, pp. 737–742.
13. Dhillon, B. S., *Medical Device Reliability and Associated Areas*, CRC Press, Boca Raton, FL, 2000.

14. Dhillon, B. S., *Patient Safety: An Engineering Approach*, CRC Press, Boca Raton, FL, 2012.

15. Sato, K., Case Study of Maintenance of Spot-Welding Robots, *Plant Maintenance*, Vol. 14, 1982, pp. 28–29.

16. Sugimoto, N. and Kawaguchi, K., Fault Tree Analysis of Hazards Created by Robots, *Proceedings of the 13th International Symposium on Industrial Robots*, 1983, pp. 9.13–9.28.

17. Khodabandehloo, K., Duggan, F. and Husband, T. M., Reliability of Industrial Robots: A Safety Viewpoint, *Proceedings of the 7th British Robot Association Annual Conference*, 1984, pp. 233–242.

18. Khodabandehloo, K., Duggan, F. and Husband, T. M., Reliability Assessment of Industrial Robots, *Proceedings of the 14th International Symposium on Industrial Robots*, 1984, pp. 209–220.

19. Herrmann, D. S., *Software Safety and Reliability*, IEEE Computer Society Press, Los Alamitos, CA, 1999.

20. Dhillon, B. S., *Design Reliability: Fundamentals and Applications*, CRC Press, Boca Raton, FL, 1999.

21. Dhillon, B. S., *Human Reliability: With Human Factors*, Pergamon Press, New York, 1986.

22. Young, J. F., *Robotics*, Butterworth, London, 1973.

23. MIL-HDBK-217, *Reliability Prediction of Electronic Equipment*, Department of Defense, Washington, DC.

24. Palady, P., *Failure Modes and Effects Analysis*, PT Publications, West Palm Beach, FL, 1995.

25. Coutinho, J.S., Failure Effect Analysis, *Transactions of the New York Academy of Sciences*, Vol. 26, Series II, 1963–1964, pp. 564–584.

26. *Fault Tree Handbook*, Report No. NUREG-0492, U.S. Nuclear Regulatory Commission, Washington, DC, 1981.

27. Dhillon, B. S. and Singh, C., *Engineering Reliability: New Techniques and Applications*, John Wiley & Sons, New York, 1981.

28. Shooman, M. L., *Probabilistic Reliability: An Engineering Approach*, McGraw-Hill, New York, 1968.

29. Dhillon, B. S. and Li, Z., Stochastic Analysis of a System with Redundant Robots, One Built-in Safety Unit, and Common Cause Failures, *Journal of Intelligent and Robotics Systems*, Vol. 45, 2006, pp. 137–155.

30. Dhillon, B. S. and Fashandi, A. R. M., Robotic Systems Probabilistic Analysis, *Microelectronics and Reliability*, Vol. 37, 1997, pp. 211–224.

31. Dhillon, B. S. and Fashandi, A. R. M., Stochastic Analysis of a Robot Machine with Duplicate Safety Units, *Journal of Quality in Maintenance Engineering*, Vol. 5, No. 2, 1999, pp. 114–127.

32. Allen, D., California Home to Almost One-Fifth of U.S. Medical Device Industry, *Medical Device and Diagnostics Magazine*, Vol. 19, No. 10, 1997, pp. 64–67.

33. Micco, L. A., Motivation for the Biomedical Instrument Manufacturer, *Proceedings of the Annual Reliability and Maintainability Symposium*, 1972, pp. 242–244.

34. Walter, C. W., Instrumentation Failure Fatalities, *Electronics News*, January 27, 1969.

35. Dhillon, B. S., Reliability Technology in Health Care Systems, *Proceedings of the IASTED International Symposium on Computers Advanced Technology in Medicine, Health Care, Bio-Engineering*, 1990, pp. 84–87.
36. Schwartz, A. P., A Call for Real Added Value, *Medical Industry Executive*, February/March 1994, pp. 5–9.
37. Banta, H. D., The Regulation of Medical Devices, *Preventive Medicine*, Vol. 19, 1990, pp. 693–699.
38. *Medical Devices, Hearings Before the Subcommittee on Public Health Environment*, U. S. Congress House Interstate and Foreign Commerce, Serial No. 93-61, U.S. Government Printing Office, Washington, D. C., 1973.
39. Rose, H. B., A Small Instrument Manufacturer's Experience with Medical Equipment Reliability, *Proceedings of the Annual Reliability and Maintainability Symposium*, 1972, pp. 251–254.
40. Kohn, L. T., Corrigan, J. M., Donaldson, M. S., *To Err is Human: Building a Safer Health System, Institute of Medicine Report*, National Academy Press, Washington, D.C., 1999.
41. Bogner, M. S., Medical Devices, A New Frontier for Human Factors, *CSERIAC Gateway*, Vol. 4, No. 1, 1993, pp. 12–14.
42. Sawyer, D., *Do It By Design: Introduction to Human Factors in Medical Devices*, Center for Medical Devices and Radiological Health (CDRH), Food and Drug Administration, Washington, D.C., 1996.
43. Novel, J. L., Medical Device Failures and Adverse Effects, *Pediatric Emergency Care*, Vol. 7, 1991, pp. 120–123.
44. Bogner, M. S., Medical Devices and Human Error, in *Human Performance in Automated Systems: Current Research and Trends*, edited by M. Mouloua and R. Parasuraman, Lawrence Erlbaum Associates, Hillsdale, NJ, 1994, pp. 64–67.
45. Casey, S., *Set Phasers on Stun: And Other True Tales of Design Technology and Human Error*, Aegean Inc., Santa Barbara, CA, 1993.
46. Hyman, W. H., Human Factors in Medical Devices, in *Encyclopaedia of Medical Devices and Instrumentation*, edited by J.G. Webster, Vol. 3, John Wiley & Sons, New York, 1988, pp. 1542–1553.
47. Wikland, M. E., *Medical Device and Equipment Design*, Interpharm Press Inc., Buffalo Grove, IL, 1995.
48. Cohen, T., Validating Medical Equipment Repair and Maintenance Metrics: A Progress Report, *Biomedical Instrumentation and Technology*, Jan./Feb. 1997, pp. 23–32.
49. Dhillon, B. S., *Medical Device Reliability and Associated Areas*, CRC Press, Boca Raton, FL, 2000.
50. Taylor, E. F., The Reliability Engineer in the Health Care System, *Proceedings of the Annual Reliability and Maintainability Symposium*, 1972, pp. 245–248.
51. Center for Devices and Radiological Health (CDRH), Food and Drug Administration (FDA), 1390 Piccard Drive, Rockville, MD 20850, USA.
52. Emergency Care Research Institute (ECRI), 5200 Butler Parkway, Plymouth Meeting, PA 19462.

chapter six

Transportation system failures and oil and gas industry equipment reliability

6.1 Introduction

Each year, a vast sum of money is spent globally to develop, manufacture, and operate transportation systems such as motor vehicles, trains, ships, and aircraft. These systems carry billions of passengers and billions of tons of goods annually from one point to another around the globe. For example, as per the International Air Transport Association (IATA), the world's airlines alone carry over 1.6 billion passengers for leisure and business travel, and over 40% of the world's trade of goods is carried by air each year [1,2].

Needless to say, transportation system failures nowadays have become an important issue because they can, directly or indirectly, affect the world economy and the environment, as well as transportation reliability and safety.

Billions of dollars are spent worldwide every year to manufacture, operate, and maintain various types of equipment for the oil and gas industry. Nowadays, the reliability of equipment used in the oil and gas industrial sector has become a very important issue due to various types of equipment reliability–related problems. For example, in 1996, the direct cost of corrosion-related failure, including maintenance, in the U.S. petroleum industry alone was $3.7 billion per year [3,4].

Needless to say, equipment being used in the oil and gas industrial sector is becoming more sophisticated and complex, and its cost is increasing quite rapidly. To meet production-related targets, oil and gas organizations and companies are increasingly demanding better equipment reliability.

This chapter presents several important aspects of transportation system failure and oil and gas industry equipment reliability.

6.2   Mechanical failure–related aviation accidents

Over the years, there have been a significant number of aviation accidents that were directly or indirectly due to mechanical failures and mechanical-related pilot errors (in this case, a mechanical-related pilot error is defined as one in which pilot error was the actual cause, but it occurred due to some kind of mechanical failure). A worldwide study of 1,300 fatal accidents during the period 1950–2008 involving commercial aircraft (i.e., excluding helicopters and aircraft with 10 or fewer people on board) revealed that 25 accidents occurred due to mechanical-related pilot error and 134 accidents were caused by mechanical failure [5].

Seven of the aviation accidents that occurred due to mechanical failure are briefly described here [2]:

1. *Turkish Airlines Flight 981 accident*: This incident involved Turkish Airlines Flight 981 (aircraft type: McDonald Douglas DC-10-10), a flight from Istanbul to Heathrow Airport, London, via Paris that took place on March 3, 1974 [2,6]. The flight crashed due to cargo hatch failure and control cable failures, resulting in the deaths of all 346 people aboard.

2. *US Air Flight 427 accident*: This incident involved US Air Flight 427 (aircraft type: Boeing 737–387), a flight from Chicago's O'Hare Airport to West Palm Beach, Florida, via Pittsburgh on September 8, 1994 [2,7]. The flight crashed due to rudder device malfunction and resulted in 132 fatalities.

3. *British International Helicopters Chinook Accident*: This event involved a Boeing 234LR Chinook helicopter operated by British International Helicopters, and it occurred on November 6, 1986 [2,8]. As it approached to land at Sumburgh Airport, Shetland Islands in the United Kingdom, the helicopter crashed into the sea and sank because of a malfunction of the modified level ring gear in the forward transmission. The accident resulted in 45 fatalities and 2 injuries.

4. *British Overseas Airways Corporation (BOAC) Flight 781 accident*: This incident involved BOAC Flight 781 (aircraft type: de Havilland DH-106 Comet 1), a flight from Kallang Airport, Singapore, to Heathrow Airport, London, via Rome, and it occurred on January 10, 1954 [2,9]. The flight crashed into the Mediterranean Sea due to failure of the cabin pressure because of metal fatigue caused by the repeated depressurization and pressurization of the aircraft cabin, and 35 fatalities resulted.

5. *United Airlines Flight 585 accident*: This event involved United Airlines Flight 585 (aircraft type: Boeing 737–291), a flight from Stapleton

International Airport, Denver, to Colorado Springs, on March 3, 1991 [2,10]. The flight crashed due to rudder device malfunction and caused 25 fatalities.

6. *Los Angeles Airways Flight 841 accident*: This incident involved Los Angeles Airways Flight 841(aircraft type: Sikorsky S-61L helicopter), a flight from Disneyland Heliport, Anaheim, California, to Los Angeles International Airport on May 22, 1968 [2,11]. The flight crashed because of a mechanical failure in the blade rotor system and resulted in the deaths of all 23 people on board.

7. *United Airlines Flight 859 accident*: This incident involves United Airlines Flight 859 (aircraft type: Douglas DC-8-20), a flight from Omaha, Nebraska, to Stapleton International Airport, Denver, on July 11, 1961 [2,12]. The aircraft crashed while landing at the airport because it suffered hydraulic failure en route. The accident caused 18 fatalities and 84 injuries.

6.3 Defects in vehicle parts and classifications of vehicle failures

A motor vehicle is composed of many subsystems and parts such as the brakes, steering, rim, engine, clutch, and transmission [2,13]. Defects in such subsystems and parts can result in motor vehicle failure. Thus, defects in a number of these automobile parts are discussed next [2,13,14].

• *Brake defects*: In normal driving environments, the failure of parts in the motor vehicle's braking system is likely to take place only when the parts become defective, severely worn, or degraded. Nonetheless, the brake defects may be grouped into four categories: drum brake system defects, disk brake system defects, common disk and drum brake systems defects, and air brake system defects.

Defects in the drum brake system include brake jams, brake imbalance, the brake pedal touching the floor, increasing heat in the brakes while driving the vehicle, low braking performance and hard pedal, and noise generation during braking. Some of the defects belonging to the disk brake system defects category are excessive wear of the pad, excessive brake pedal travel, and low or no brake force.

Common disk and drum brake systems defects include excessive pedal force, brake pedal vibrations, brake fade, and a soft pedal. Finally, some of the defects belonging to the air brake system defects category are slow pressure buildup in the reservoir, no or low brake force, and slow brake response or no release.

- *Steering system defects*: These defects can result in severe motor vehicle accidents. The causes for steering system defects include poor maintenance, faulty changes made to the steering system, faulty design, faulty manufacturing, and inadequate inspection.
- *Rim defects*: These defects are as important as defects in any other portion of a motor vehicle because they can result in very serious accidents. As per Reference 13, 1,300–2,000 truck tire–related failures result in an accident, and the U.S. Bureau of Motor Carrier Safety findings indicate that around 7%–13% of all tractors and trailers have experienced at least one defective tire [2,13]. The causes of rim defects include abusive operation, poor design, and faulty manufacturing operations.

Failures of vehicles carrying passengers can be grouped into the following four classifications [2,15]:

1. *Classification 1*: In this case, the vehicle stops or is required to stop and is pushed or towed by an adjacent vehicle to the nearest station. At this point, all personnel in both the affected vehicles egress, and the failed vehicle is pushed/towed to necessary maintenance.
2. *Classification 2*: In this case, the vehicle stops and it cannot be either pushed or towed by an adjacent vehicle, and it must wait for a rescue vehicle.
3. *Classification 3*: In this case, the involved vehicle is required to reduce its speed, but it is allowed to continue to the nearest station, where all its passengers must exit, and then it is dispatched for necessary maintenance.
4. *Classification 4*: In this case, the involved vehicle is allowed to continue to the nearest station, where all its passengers must exit, and then it is dispatched for necessary maintenance.

6.4   Rail defects and weld failures

Although rails' wear resistance is basically controlled by hardness, it is also very dependent on the stresses that rails are subject to. These stresses, which include contact stresses, bending stresses, thermal stresses, and residual stresses, control the development of rail defects that can eventually result in failure [2,16,17].

Contact stresses originate from the wheel load, traction, and braking and steering-related actions. Bending stresses act either vertically or laterally, and the vertical ones are generally compressive in the rail head and tensile in the rail base. Thermal stresses originate from welding processes

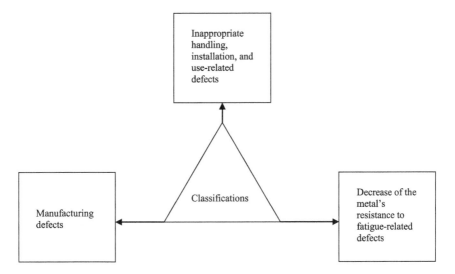

Figure 6.1   Classifications of the defects in steel rails.

during the connection of rail sections while creating a continuously welded rail, whereas residual stresses originate from manufacturing processes.

Defects in steel rails may be grouped in the three classifications shown in Figure 6.1 [2,17].

The manufacturing defects originate from the rail manufacturing process. The inappropriate handling, installation, and use-related defects originate from out-of-specification installation of rails, wheel burns, and unexpected scratches. Finally, the decrease of the metal's resistance to fatigue-related defects include the most common rail defects, such as squats and head checks.

Various methods can be used to detect rail defects to reduce the occurrence of rail defect–related failures. These methods include ultrasonic defect detection, impedance spectroscopy, visual inspection by the track maintenance staff, electromagnetic acoustic transducers, and eddy-current testing [16,17].

In railway systems, the construction of continuous welded rail (CWR) is indispensable to reducing the cost of track maintenance, improving the ride quality, and reducing noise and vibration. Over the years, many railway accidents have occurred because of weld failures. Therefore, it is extremely important to highly reliable welds for eliminating the occurrence of weld failures in service, as well as for extending the CWR service life.

Past experiences over the years clearly indicate that the majority of rail weld failures begin from weld discontinuities, and fusion welding tends to cause such weld discontinuities easily [18]. Therefore, fusion-welding methods such as aluminothermic welding (TW) and enclosed-arc (EA) welding are less reliable than pressure-welding methods such as gas-pressure (GP) welding and flash welding (FW) [2,18,19]. Thus, to eliminate the occurrence of rail weld-related failures, it is extremely important to carry out reliable welding by utilizing appropriate welding processes, welding conditions, adequate inspection methods, and well-trained welding technicians.

6.5   Road and rail tanker failure modes failure consequences

Road and rail tankers are used for carrying liquefied gases and other hazardous liquids from one location to another. Over the years, the malfunction of such tankers has led to very serious consequences. The main locations of these malfunctions are pumps, inspection covers, shells, valves, connections to a container, and branches, including instrument connections.

Road and rail tanker failure modes may categorized under the following classifications [2,20]:

- *Classification I: Metallurgical failures*—The main causes of these failures include erosion, vessel used for a purpose not covered by specifications, fatigue, use of wrong or inadequate construction materials, corrosion (external or internal), embrittlement by chemical action, vessel designed/constructed to an inadequate specification, and failure to satisfy specified construction codes.
- *Classification II: Failures due to mechanical causes besides overpressure*—The main causes of these failures include collision with another vehicle, collision with a fixed object such as a bridge, general wear and tear, damage by an external explosion, modifications in violation of original specifications, and collapse of a structure on it.
- *Classification III: Failures due to excess internal pressure*—The main causes of these failures are flame impingement, hydraulic rupture consequent upon overfilling, tanker contents having higher vapor pressure than the specifications call for, internal chemical reactions such as decomposition or polymerization, and abnormal meteorological conditions.

There are various consequences of road and rail tanker failures involving loss of containment. The nature of such consequences is influenced, directly or indirectly, by the following principal factors [2,20]:

- *Principal factor I*: The physical state of the contents
- *Principal factor II*: The nature of the surroundings

- *Principal factor III*: The location and size of any leak that develops
- *Principal factor IV*: The mechanism of dispersion
- *Principal factor V*: The chemical nature of the contents

6.6   Ship failures and their causes

The shipping industrial sector is composed of various types of ships, including carriers, tankers, container ships, and bulk cargo ships. These ships contain various types of equipment, systems, and parts that can occasionally malfunction. Some examples of these equipment, system, and part failures are shown in Figure 6.2 [2]. The consequences of these failures can vary quite considerably.

Nonetheless, the ship failures can occur due to many distinct causes. The common ones are as follows [2]:

- Fatigue
- Welding defects
- Corrosion
- Manufacturing defects
- Improper maintenance
- Inadequate quality assurance
- Unforeseen operating conditions

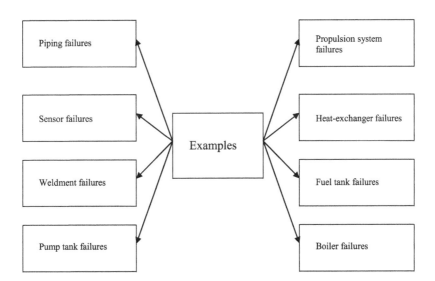

Figure 6.2   Examples of ships' equipment, system, and part failures.

6.7   *Microanalysis techinques for failure analysis*

Nowadays, modern vessels contain many polymeric components/parts, such as electrical insulation and pressure seals, and some of these are critical to the operation of the vessel. A number of microanalysis techniques that are considered quite useful in failure investigations that involve polymers. Four commonly used microanalysis techniques are described separately in the next subsections [2,21].

6.7.1   *Thermomechanical analysis*

Thermomechanical analysis involves measuring changes in length or volume of a sample as a function of time or temperature. The technique is commonly utilized for determining the thermal expansion coefficient, as well as the glass-transition temperature of polymer/composite materials. A weighted probe is placed on the specimen's surface, and the probe's vertical movement is monitored on a continuous basis while the sample is heated at a controlled rate.

6.7.2   *Thermogravimetric analysis*

Thermogravimetric analysis measures changes in the weight of the sample under consideration as a function of temperature or time. The technique is utilized for determining polymer degradation temperatures, levels of residual solvent, absorbed moisture content, and the degree of inorganic (i.e., noncombustible) filler in polymer or composite material compositions. The method can also be quite useful in the deformulation of complex polymer-based products.

6.7.3   *Differential scanning calorimetry (DSC)*

The differential scanning calorimetry technique measures heat flow to a polymer. This is very important because, by monitoring the heat flow as a function of temperature, phase transitions such as glass-transition temperatures and crystalline melt temperatures can be characterized quite effectively. This, in turn, is very useful for determining how a polymer will behave at operational temperatures.

The technique can also be utilized in forensic investigations for determining the maximum temperature to which a polymer has been subjected. This can be quite useful in establishing whether a system/equipment/part has been subjected to thermal-related overloads during service. Finally, this technique also can be utilized for determining the thermal stability of polymers by measuring the oxidation induction temperature/time.

6.7.4   Fourier transform infrared spectroscopy

Fourier transform infrared spectroscopy is utilized for identifying and characterizing polymer materials and their additives. It is a quite useful technique, particularly for highlighting defects or inclusions in plastic films or molded parts. Additional information on this method is available in Reference 21.

6.8   Mechanical seal failure

Mechanical seals have been increasingly utilized for sealing rotating shafts for more than seven decades. Nowadays, in the oil and gas industry, they are the most common types of seals found on items such as centrifugal pumps and compressors. Over the years, mechanical seal failures have become a very important issue. For example, a study carried out in a petroleum company reported that around 60% of plant breakdowns were due to mechanical seal failure, either directly or indirectly [22].

A study of mechanical seal failures in a company reported the following causes [22]:

- *Cause I*: Metal particles embedded in the carbon
- *Cause II*: Dry-running
- *Cause III*: Wrong seal spring compression
- *Cause IV*: Highly worn carbon (i.e., greater than 4 mm of wear)
- *Cause V*: Shaft and seal face plane misaligned
- *Cause VI*: Seal component failure (i.e., other than the faces or secondary seals)
- *Cause VII*: Hang-up (i.e., crystallization)
- *Cause VIII*: Auxiliary seal system failure (e.g., flush, quench, recirculating, cooling)
- *Cause IX*: Hang-up (i.e., coking)
- *Cause X*: External system or component failure (e.g., bearings)

6.8.1   Typical failure modes for mechanical seals
and their causes

Due to various causes, mechanical seals can fail in many failure modes. Nine of the typical failure modes and their corresponding causes for mechanical seals are presented here [4,23]:

- *Failure mode I: Accelerated seal face wear*—Its causes are inadequate lubrication, excessive torque, excessive shaft end play, shaft out-of-roundness, surface finish deterioration, misalignment, and contaminants.

- *Failure mode II: Seal face edge chipping*—Its causes are excessive shaft deflection, seal faces out-of-square, and excessive shaft whip.
- *Failure mode III: Seal face distortion*—Its causes are excessive pressure on a seal, foreign material trapped between faces, insufficient seal lubrication, and excessive pressure velocity (PV) value of seal operation.
- *Failure mode IV: Fractured spring*—Its causes are material flaws, misalignment, stress concentration due to tooling marks, and corrosion.
- *Failure mode V: Excessive friction resulting in slow mechanical response*— Its causes are excessive squeeze, metal-to-metal contact (i.e., being out of alignment), excessive seal swell, and seal extrusion.
- *Failure mode VI: O-ring failure*—Its causes are excessive temperature (i.e., greater than 55°C), excessive fluid pressure, and installation error.
- *Failure mode VII: Compression set and low pressure leakage*—This is caused by extreme temperature operation.
- *Failure mode VIII: Seal fracture*—Its causes are stress-corrosion cracking, excessive PV value, and excessive fluid pressure on a seal.
- *Failure mode IX: Small leakage*—Its causes are insufficient squeeze and installation damage.

6.9   Optical connector failures

Optical fiber connectors are utilized for joining optical fibers in conditions where a connect or disconnect is needed. Fiberoptic equipment, including wet-mate optical connectors, is a very important element of the current subsea infrastructure in oil and gas applications. A study of reliability data concerning optical connectors (excluding cables or jumpers) collected over a 10-year period reported four factors (including their percentage breakdowns), as shown in Figure 6.3, which caused optical connector failures, directly or indirectly [4,24].

The data include field failures, as well as failures that occurred during integration into equipment and testing process prior to field deployment.

Note that, based on the percentages shown in Figure 6.3, only 14% of optical connector failures were related to optical issues and the remaining 86% of failures were due to material, mechanical, and external factors. Furthermore, when only field failures were studied, the optical connector failures occurred due to mechanical and material issues [24]. They accounted for 61% and 39% of the failures, respectively.

Additional information on optical connector-related failures is available in Reference 24.

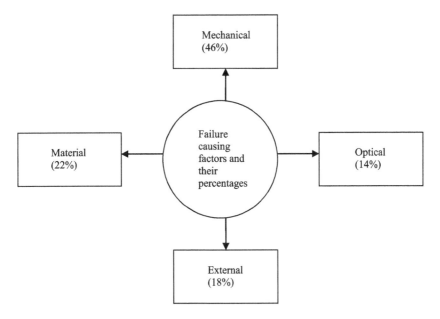

Figure 6.3   The four factors that cause optical connector failure, along with their percentages.

6.10   Corrosion-related failures

Corrosion in the oil and gas industrial sector has been acknowledged since the 1920s, and corrosion and other related failures continue to cost the off-shore oil and gas industrial sector a vast sum of money each year [25]. As per Reference 26, in the oil and gas industry, corrosion-related failures constitute more than 25% of failures. A study conducted in the 1980s reported the following causes (along with their degree of contribution, given as percentages) for corrosion-related failures in petroleum-related industries [26,27]:

- *Cause I*: Stress corrosion: 3%
- *Cause II*: Impingement: 3%
- *Cause III*: Crevice: 3%
- *Cause IV*: Galvanic: 6%
- *Cause V*: Erosion corrosion: 9%
- *Cause VI*: Pitting 12%
- *Cause VII*: Preferential weld: 18%
- *Cause VIII*: H$_2$S related: 18%
- *Cause IX*: CO$_2$ related: 28%

6.10.1 *Types of corrosion/degradation that can cause failure*

There are several possible types of corrosion/degradation that can cause failure [26]:

- *Weight loss corrosion*: This type of corrosion occurs most commonly in oil and gas production because of an electrochemical reaction between metal and the corrodents in the environment.
- *Galvanic corrosion*: This type of corrosion occurs in bimetallic connections on opposite ends of the galvanic series that have enough potential for causing a corrosion reaction in the existence of an electrolyte.
- *Corrosion fatigue*: Over the years, it has played a very important role in drilling and surface operations, such as sucker-rod and drill pipe failures.
- *Fretting corrosion*: This type of corrosion generally takes place in poorly lubricated valve stems, where a partially opened valve causes some vibration that can result in galling and then, in turn, valve seizure and possible failure.
- *Stress corrosion cracking*: In oil and gas production, the most probable form of the cracking phenomenon is sulfide and chloride stress corrosion cracking.
- *Crevice corrosion*: This type of corrosion takes place in conditions where crevice forms, such as partial penetration welds and backing strips, are utilized.
- *Microbiological-induced corrosion*: This type of corrosion is very serious, as it takes the form of localized pitting attacks that can cause rapid loss of metal in a concentrated area, leading to ruptures or leaks.
- *Impingement/cavitation*: Impingement can take place in situations where process fluid is forced to change its flow direction abruptly. For the occurrence of cavitation, the common offshore area is in pump impellors, where pressure changes take place and high liquid flow rates occur.
- *Hydrogen-induced cracking*: Past experiences over the years indicate that this type of cracking has mostly taken place in controlled rolled pipeline steel with elongated stringers of nonmetallic imperfections.
- *Erosion-corrosion*: It is often observed on the outer radius of pipe bends in oil and gas production because of high fluid flow rates, as well as corrosive environments where flow exceeds 10 m/s for carbon steel and 6 m/s for copper nickel.

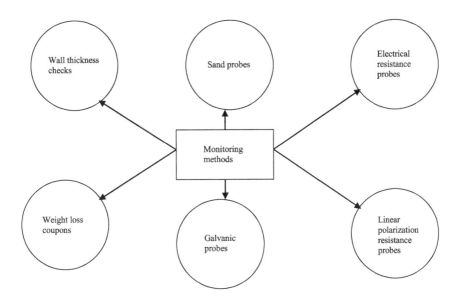

Figure 6.4   Commonly used corrosion/condition monitoring methods.

6.10.2   Corrosion- and condition-monitoring methods

Corrosion monitoring of internal surfaces may be conducted by using a combination of the five methods presented here [4,25]:

1. *Method I*: Visual inspection
2. *Method II*: Pipeline leak detection
3. *Method III*: Measurements of nonintrusive wall thickness (radiography/ultrasonic)
4. *Method IV*: Chemical analysis of samples taken from the product
5. *Method V*: Intrusive probes and coupons for monitoring corrosion and erosion

Six commonly used corrosion- and condition-monitoring methods are shown in Figure 6.4 [4,25]. Additional information on these methods is available in Reference 25.

6.11   Common cause failure defense approach for an oil and gas industry safety instrumented system (SIS)

In the oil and gas industry, a safety instrumented system (SIS) normally functions in a low demand mode, which means that regular inspection and testing are necessary for revealing its failure. Past experiences over the years clearly indicate that the occurrences of common cause failures are a serious threat to SIS reliability and may result in simultaneous failures of redundant units and parts and safety barriers [4,28–30].

Thus, a common cause failure may simply be defined as any instance where multiple units, components, or parts fail due to a single cause [30]. Some causes of common cause failures are operation and maintenance errors, common external power source, design deficiency, external catastrophe, common manufacturer, and external normal environment [30,31].

The three key aspects on which the common cause failure defense approach for oil and gas industry SIS focuses are as follows:

1. To highlight common cause failures and their causes on the basis of failure-related reports
2. To avoid introducing common cause failures during inspection and function testing–related processes
3. To use the insight of failure causes for selecting efficient mechanisms to defend against future occurrences of common cause failures

6.11.1 Common cause failure defense approach

The common cause failure defense approach is based on six function testing and inspection tasks. In turn, these tasks are based on checklists and analytical methods such as influence diagrams, operational sequence diagrams (OSDs), and cause-defense matrices. The tasks are scheduling; preparation, execution, and restoration; failure reporting; failure analysis; implementation; and validation and continuous improvements.

All these six tasks are described in the next subsections [4,29].

6.11.1.1 Task 1: Scheduling

Task 1 is concerned with ensuring that during the scheduling process, all appropriate improvements are captured. Here, note that a very important defense against the occurrence of common cause failure is to ensure that any corrections and improvements to the test procedure are properly captured during the creation of a new function test or inspection work packages.

6.11.1.2 Task 2: Preparation, execution, and restoration

Task 2 is concerned with avoiding the introduction of common cause failures during preparation, execution, and restoration processes. Three checklists, containing questions for preparation, execution, and restoration processes, are considered very useful [4,29]:

1. Preparation checklist questions:
   a. Are all the calibration tools properly calibrated?
   b. Are all the individuals involved in executing the test clearly familiar with calibration and testing tools?
   c. Have compensating appropriate measures been properly highlighted and implemented for avoiding human errors?
   d. Have all types of potential human errors during execution and restoration been clearly discussed and highlighted?
   e. Have all the human error–related incidents been experienced during the earlier execution?
   f. Does the procedure describe essential steps for safely restoring the SIS?
   g. Does the procedure contain all known deficiencies (e.g., ambiguous instructions)?
2. Execution checklist questions:
   a. Are all the process connections free of plugging and (if applicable) heat-traced?
   b. Are all the field SIS parts (constituting the safety function under test) clearly labeled?
   c. Are all the additional parts that are operated during SIS function testing and inspection process appropriately labeled?
   d. Are all the parts appropriately protected against damage from nearby work-related activities?
   e. Are all the parts operated within the stated operating and environmental conditions?
3. Restoration checklist questions:
   a. Are any remaining bypasses, inhibits, or overrides logged, and compensating measures appropriately highlighted and implemented?
   b. Has the safety function been appropriately verified prior to startup?
   c. Have the suspensions of all inhibits and overrides been appropriately verified and communicated?
   d. Has the physical restoration (e.g., bypasses and isolation valves) been appropriately verified?

6.11.1.3   Task 3: Failure reporting

Task 3 is concerned with improving the quality of failure reporting. In this context, the six questions presented here are considered very useful [4,29]:

1. How was the failure uncovered or observed (i.e., during the inspection or repair process, by diagnostic, by review, incidentally, upon demand, or during function testing)?
2. What appears to be the failure causes?

3. Was the part tested or inspected in a different way than stated in the inspection or test procedure; if so, what was the reason for the approach to be different?
4. What was the effect of the failure's occurrence on the overall safety function (i.e., degraded, loss of entire function, none at all)?
5. Has the part or component been overexposed (i.e., environmental or by operational stresses); if so, what could the associated causes be?
6. Have similar types of failure occurred previously?

6.11.1.4　Task 4: Failure analysis

Task 4 is concerned with highlighting common cause failures through failure analysis. The four steps presented here are considered very useful in highlighting common cause failures [4,29]:

1. Step 1: Review the failure's description and verify the initial failure classification (and, if necessary, correct it).
2. Step 2: Carry out an appropriate initial screening that captures failures that (1) clearly share failure-associated causes, (2) have been found within the framework of the same test or inspection interval, (3) have very similar physical location or design, and (4) the causes for failures are not random, as stated in IEC 61508, 1998, and 61511m 2003 documents [29].
3. Step 3: Carry out a root cause and coupling factor analysis with the aid of influence diagrams.
4. Step 4: List all the root cause and coupling-related factors in a cause-defense matrix.

6.11.1.5　Task 5: Implementation

Task 5 is concerned with implementing appropriate defensive measures. The proper implementation of common cause failure–related defensive measures is extremely important for preventing the occurrence of similar types of failures. Additional information on this task is available in Reference 29.

6.11.1.6　Task 6: Validation and continuous improvements

Task 6, the final task, is concerned with performing validation and continuous improvements. With regard to validation, the questions presented here are considered quite useful [4,29]:

1. Are all the common cause failures systematically highlighted and analyzed and appropriate defenses implemented for preventing their occurrence?

2. Are all the failures introduced during inspection and testing processes captured, analyzed, and utilized for improving the related procedures?
3. Are all disciplines concerned with SIS inspection, testing, maintenance, and follow-up clearly familiar with the common cause failures concept?
4. Are all personnel utilizing the test and calibration tools clearly familiar with proper application?
5. Are all the failures detected upon real demands appropriately analyzed for verifying that they would have been detected during an inspection/function test?
6. Are all the dangerous undetected failure modes clearly known and appropriately catered to in the function testing and inspection-associated procedures?
7. Are all changes in operating or environmental conditions appropriately captured and analyzed for essential modifications to the SIS or related procedures?
8. Are all the procedure-associated shortcomings appropriately communicated to all the responsible individuals and followed up properly?
9. Are all the calibration and test tools clearly suitable and maintained as per the vendor recommendations?
10. Are all the requirements for the safety function appropriately covered by the inspection or function test procedures?
11. Are all the test-associated limitations (compared to the actual demand conditions) clearly known?
12. Are all the diagnostic alarms appropriately followed up on within the started mean time to restoration?
13. Are all safety function redundant channels appropriately covered by the function testing or inspection-associated procedures?

6.12   Oil and gas pipeline fault tree analysis

The fault tree method was developed in the early 1960s at the Bell Telephone Laboratories for the purpose of performing analysis of the Minuteman Launch Control System with regard to safety [31]. Nowadays, the method is used around the globe for conducting various types of reliability and safety-related studies. This method is described in detail in Chapter 4. The application of this method for performing oil-gas long pipeline failure analysis is demonstrated through the two examples presented next [4,32].

EXAMPLE 6.1

Assume that the failure of an oil-gas pipeline can occur due to any of the following events: misoperation, third-party damage, pipeline with defects, material with poor mechanical properties, and pipeline with serious corrosion. The first three of these events are described here:

1. The event "misoperation" can occur due to operate misoperation, design misoperation, or maintain operation.
2. The event "third-party damage" can occur due to either a natural disaster and external force or artificial damage.
3. The event "pipeline with defects" can occur due to a pipeline with either initial defects or construction defects.

Using the fault tree symbols given in Chapter 4, develop a fault tree for the top event, "Oil-gas pipeline failure."

A fault tree for this example is shown in Figure 6.5. The single capital letters in the rectangles and circles in the drawing denote the corresponding fault events (i.e., T: Oil-gas pipeline failure, A: Pipeline with serious corrosion, B: Material with poor mechanical properties, C: Misoperation, D: Third-party damage, E: Pipeline with defects, F: Pipeline with initial defects, G: Pipeline with construction defects, H: Natural disaster and external force, I: Artificial damage, K: Maintain misoperation, L: Operate misoperation, and M: Design misoperation).

EXAMPLE 6.2

Assume that in Figure 6.5, the probabilities of the fault events A, B, H, I, F, G, K, L, and M are 0.01, 0.02, 0.03, 0.04, 0.05, 0.06, 0.07, 0.08, and 0.09,

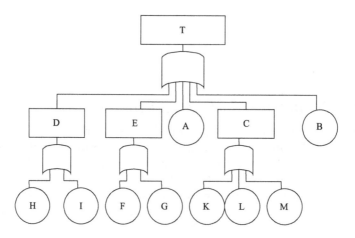

Figure 6.5   A fault tree for the top event T: Oil-gas pipeline failure.

respectively. With the aid of Chapter 4, calculate the probability of occurrence of the top fault event T: Oil-gas pipeline failure.

By using the material presented in Chapter 4 and the fault events' occurrence probabilities, we calculate the probabilities of fault events D, E, C, and T as follows:

The probability of the occurrence of fault event D is

$$
\begin{aligned}
P(D) &= 1 - (1 - P(H))(1 - P(I)) \\
&= 1 - (1 - 0.03)(1 - 0.04) \\
&= 0.0688,
\end{aligned} \tag{6.1}
$$

where $P(H)$ is the occurrence probability of fault event H and $P(I)$ is the occurrence probability of fault event I.

Similarly, the probability of the occurrence of fault event E is

$$
\begin{aligned}
P(E) &= 1 - (1 - P(F))(1 - P(G)) \\
&= 1 - (1 - 0.05)(1 - 0.06) \\
&= 0.107,
\end{aligned} \tag{6.2}
$$

where $P(F)$ is the occurrence probability of fault event F and $P(G)$ is the occurrence probability of fault event G.

Similarly, the probability of the occurrence of fault event C is

$$
\begin{aligned}
P(C) &= 1 - (1 - P(K))(1 - P(L))(1 - P(M)) \\
&= 1 - (1 - 0.07)(1 - 0.08)(1 - 0.09) \\
&= 0.2214,
\end{aligned} \tag{6.3}
$$

where $P(K)$ is the occurrence probability of fault event K, $P(L)$ is the occurrence probability of fault event L, and $P(M)$ is the occurrence probability of fault event M.

Finally, by using these calculated fault events' occurrence probability values, we get

$$
\begin{aligned}
P(T) &= 1 - (1 - P(A))(1 - P(D))(1 - P(E))(1 - P(C))(1 - P(B)) \\
&= 1 - (1 - 0.01)(1 - 0.0688)(1 - 0.107)(1 - 0.2214)(1 - 0.02) \\
&= 0.6281,
\end{aligned}
$$

$$\tag{6.4}$$

where $P(T)$ is the occurrence probability of the top fault event T (i.e., Oil-gas pipeline failure).

Thus, the probability of occurrence of the top fault event T (Oil-gas pipeline failure) is 0.6281. A redrawing of the fault tree in Figure 6.5 with the given and calculated fault events' occurrence probability values is shown in Figure 6.6.

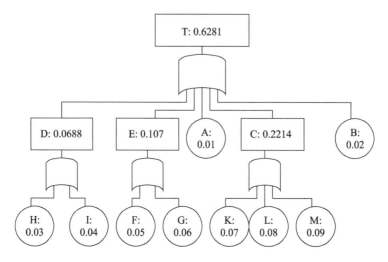

Figure 6.6 The fault tree in Figure 6.5, redrawn with given and calculated fault events' occurrence probability values.

PROBLEMS

1. Discuss the following mechanical failure–related aviation accidents:
   i. BOAC Flight 782 accident
   ii. Turkish Airlines Flight 981 accident
   iii. United Airlines Flight 585 accident
2. Describe the following defects associated with a vehicle:
   i. Rim defects
   ii. Brake defects
3. Discuss the classifications of defects in steel rails.
4. Discuss ship failures and their causes.
5. Describe the following microanalysis techniques for failure analysis:
   i. Thermogravimetric analysis
   ii. Thermomechanical analysis
6. Discuss optical connector failures.
7. Discuss at least eight typical failure modes and their causes of mechanical seals.
8. What are the types of corrosion and degradation that can cause failure?
9. What are the commonly used corrosion- and condition-monitoring methods?
10. Describe the common cause failure defense approach for oil and gas industry safety instrumented systems.

References

1. *Fast Facts: The Air Transport Industry in Europe Has United to Present Its Key Facts and Figures*, International Air Transport Association (IATA), Montreal, Canada, July 2006. Available online at www.iata.org/pressroom/economics.facts/stats/2003-04-10-01.htm, accessed on March 10, 2008.
2. Dhillon, B. S., *Transportation Systems Reliability and Safety*, CRC Press, Boca Raton, FL, 2011.
3. Kane, A. D., Corrosion in Petroleum Refining and Petrochemical Operations, in *Metals Handbook, Vol. 13C: Environments and Industries*, edited by S. O. Cramer and B. S. Covino, ASM International, Metals Park, OH, 2003, pp. 967–1014.
4. Dhillon, B. S., *Safety and Reliability in the Oil and Gas Industry: A Practical Approach*, CRC Press, Boca Raton, FL, 2016.
5. *Causes of Fatal Aviation Accidents by Decade*, 2009. Available online at www.planecrashinfo.com/cause.htm, accessed on June 12, 2010.
6. Johnston, M., *The Last Nine Minutes: The Story of Flight 981*, Morrow Publishers, New York, 1976.
7. Byrne, G., *Flight 427: Anatomy of an Air Disaster*, Springer-Verlag, New York, 2002.
8. *Report on the Accident to Boeing Vertol (BV) 234LR, G-BWFC 2.5 Miles East of Sumburgh, Shetland Isles, November 6, 1986*, Report No. 2, Air Accidents Investigation Branch (AAIB), Aldershot, UK, 1988.
9. Stewart, S., *Air Disasters*, Arrow Books, London, 1986.
10. *Aircraft Accident Report: United Airlines Flight 585*, Report No. AAR92-06, National Transportation Safety Board (NTSB), Washington, DC, 1992. Available online at http://libraryonline.erau.edu/online-full-text/ntsb/aircraft-accident-reports/AAR92-06.pdf
11. Gero, D., *Aviation Disasters*, Patrick Stephens Ltd, Sparkford, UK, 1993.
12. United Airlines Flight 859, Aircraft Accident Report No. SA-362 (file 1-0003), Civil Aeronautics Board (CAB), Washington, DC, 1962.
13. Limpert, R., *Vehicle System Components: Design and Safety*, John Wiley & Sons, New York, 1982.
14. Dhillon, B. S., *Mechanical Reliability: Theory, Models and Applications*, American Institute of Aeronautics and Astronautics, Washington, DC, 1988.
15. Anderson, J. E., *Transit Systems Theory*, D.C. Heath, Lexington, MA, 1978.
16. Cannon, D. F., Edel, K. O., Grassie, S. L., and Sawley, K., Rail Defects: An Overview, *Fatigue and Fracture of Engineering Materials and Structures*, Vol. 26, No. 10, 2003, pp. 865–886.
17. Labropoulos, K. C., Moundoulas, P., and Moropoulou, A., Methodology for the Monitoring, Control, and Warning of Defects for Preventive Maintenance of Rails, *Computers in Railways X, WIT Transactions on the Built Environment*, Vol. 88, WIT Press, London, 2006, pp. 935–944.
18. Fukada, Y., Yamamoto, R., Harasawa, H., and Nakanowatari, H., Experience in Maintaining Rail Track in Japan, *Welding in the World*, Vol. 47, 2003, pp. 123–137.

19. Tatasumi, M., Fukada, Y., Veyama, K., Shitrara, H., and Yamamoto, R., Quality Evaluation Methods for Rail Welds in Japan, *Proceedings of the World Congress on Railway Research*, 1997, pp. 197–205.
20. Marshall, V. C., Modes and Consequences of the Failure of Road and Rail Tankers Carrying Liquefied Gases and Other Hazardous Liquids, in *Reliability on the Move*, edited by G. B. Guy, Elsevier Science, London, 1989, pp. 136–148.
21. Dean, R. J., *Investigation of Failures in Marine Environment*, ERA Technology Ltd, Leatherhead, Surrey, UK, 2009.
22. Wilson, B., Mechanical Seals, *Industrial Lubrication and Tribology*, Vol. 47, No. 2, 1995, p. 4.
23. Skewis, W. H. *Mechanical Seal Failure Modes*, Support Systems Technology Corporation, Gaithersburg, MD, retrieved on May 28, 2015. Available online at http://docslide.us/documents/mechanical-seal-failure-modes.html.
24. Jones, R. T., and Thiraviam, A., Reliability of Fiber Optic Connectors, *Proceedings of the IEEE OCEANS Conference*, 2010, pp. 1–10.
25. Price, J. C., Fitness-for-Purpose Failure and Corrosion Control Management in Offshore Oil and Gas Development, *Proceedings of the 11th International Offshore and Polar Engineering Conference*, 2001, pp. 234–241.
26. Kermani, M. B., and Harrop, D., The Impact of Corrosion on the Oil and Gas Industry, *Society of Petroleum Engineers (SPE) Production and Facilities*, August 1996, pp. 186–190.
27. Kermani, M. B., Hydrogen Cracking and Its Mitigation in the Petroleum Industry, *Proceedings of the Conference on Hydrogen Transport and Cracking in Metals*, 1994, pp. 1–8.
28. Summers, A. E., and Raney, G., Common Cause and Common Sense, Designing Failure Out of Your Safety Instrumented System (SIS), *ISA Transactions*, Vol. 38, 1999, pp. 291–299.
29. Lundteigen, M. A., and Rausand, M., Common Cause Failures in Safety Instrumented Systems on Oil and Gas Installations: Implementing Defense Measures Through Testing, *Journal of Loss Prevention in the Process Industries*, Vol. 20, 2007, pp. 218–229.
30. Dhillon, B. S., and Proctor, C. L., Common-Mode Failure Analysis of Reliability Networks, *Proceedings of the Annual Reliability and Maintainability Symposium*, 1977, pp. 404–408.
31. Dhillon, B. S., and Singh, C., *Engineering Reliability: New Techniques and Applications*, John Wiley & Sons, New York, 1981.
32. Tian, H. et al., Application of Fault Tree Analysis in the Reliability Analysis of Oil-Gas Long Pipeline, *Proceedings of the International Conference on Pipelines and Trenchless Technology*, 2013, pp. 1436–1446.

chapter seven

Computer system and Internet reliability and software bugs in computer systems

7.1 Introduction

Applications of computers over the years have increased at an alarming rate, ranging from personal use to controlling various types of highly sophisticated systems in business. As failures of computers can affect our day-to-day life, whether directly or indirectly, computer reliability has become a very important issue to the population at large. Furthermore, the reliability of computer systems utilized in areas such as defense, aerospace, and nuclear power generation is vital because malfunctions in these areas could be catastrophic, both financially and in terms of human cost.

The history of the Internet goes back to the late 1960s, with the development of the Advanced Research Projects Agency Network (ARPANET) [1,2]. The Internet has grown from just 4 hosts in 1969 to around 38 million sites and 147 hosts in 2002, and nowadays billions of people in the world use Web services [1–3]. Today, the reliability of the Internet has become extremely important to the world's economy and other facets of life because Internet-related failures can result in millions of dollars in losses and interrupt the day-to-day routines of a very large number of its users in the world [2,4].

Software bugs in computer systems have become a pressing issue due to problems that they cause, such as high cost and catastrophic failures. For example, in 2002, a study commissioned by the U.S. Department of Commerce's National Institute of Standards and Technology (NIST) revealed that software bugs and errors alone cost the U.S. economy approximately $59 billion per year—about 0.6% of its gross domestic product (GDP)] [5].

This chapter discusses various important aspects of computer system and Internet reliability and software bugs in computer systems.

7.2 Factors related to computer system reliability issues and sources of computer failure

There are many factors related to computer system reliability issues, including the following [2,5–7]:

- Computer-associated failures are quite highly varied in character. For example, a computer system part or component may malfunction permanently or experience a transient fault because of its environment.
- It could be extremely difficult to detect properly hardware design–related errors at the lowest system levels prior to the production and installation phases. Therefore, it is quite possible that hardware design–related errors, directly or indirectly, may lead to situations where it is impossible to distinguish among operation errors due to such oversights from the ones due to transient physical faults.
- The logic elements are the computers' main parts and components that have troublesome reliability-associated features. In many cases, the proper determination of the reliability of such elements is impossible, and their defects cannot be healed appropriately.
- Generally, dynamic fault tolerance is the most powerful type of self-repair in computers, but it is cumbersome to analyze. Nevertheless, for certain applications, it is highly important and cannot simply be overlooked.
- For fault tolerance, computers nowadays are composed of redundancy schemes; advances made over the years have brought a number of improvements, but there are still many practical and theoretical difficulties that still must be overcome.

There are many sources that, directly or indirectly, lead to computer failures. These sources include peripheral device failure, environmental and power failures, communication network failure, processor and memory failures, mysterious failures, human error, gradual erosion of the database, and saturation [2,5,8,9]. The first six of these sources are described below.

1. *Peripheral device failure*: This type of failure is important, but such instances rarely cause a system shutdown. In peripheral devices, frequently occurring errors are intermittent or transient, and the normal reason for their occurrence is the electromechanical nature of peripheral devices.
2. *Environmental and power failures*: Environmental failures take place due to causes such as electromagnetic interference, air conditioning equipment failure, and fires. The causes of power failures

include items such as total power loss from the local utility company and transient fluctuations in voltage or frequency.

3. *Communication network failure*: This type of failure is concerned with intermodule communication, and most of these issues are generally of a transient nature. The employment of vertical parity logic can help one to detect approximately two-thirds of errors in communication lines.

4. *Processor and memory failures*: Processor failures and errors are quite catastrophic, but their occurrence is rare, as there are times when the central processor malfunctions to execute instructions correctly due to a dropped bit. Nowadays, the incidence of memory parity errors is quite rare because of improvements in hardware reliability, and they are not necessarily fatal.

5. *Mysterious failures*: These failures take place unexpectedly; thus, in real-life systems, such incidents are never classified or categorized appropriately. For example, when a normally functioning system suddenly stops functioning without indicating any problem (i.e., with hardware, software or other elements), the failure is referred to as a *mysterious failure.*

6. *Human error*: This kind of error generally occurs due to operator-related oversights and mistakes. Quite often, operator errors take place during starting up, running, and shutting down the computer.

7.3 Computer-related fault catagories and reliability measures

Normally, for computer system reliability modeling evaluation, it is a good approach to categorize computer-related faults based on their duration. Thus, computer-related faults may be categorized in one of the following classifications [2,5,10]:

- *Classification I: Transient faults*—These faults occur due to the temporary failure of components or parts or external interference, such as electrical noise, power dips, and glitches. They are of limited duration, and although they need restoration, they do not involve any replacement or repair. This type of fault is characterized by the arrival modes, as well as the duration of transients [2,5,10].

- *Classification II: Permanent faults*—These faults often occur due to catastrophic failures of components and parts. In this situation, the parts' or components' failures are permanent and irreversible and need replacement or repair. These faults have a failure rate that very much depends on the surrounding environment and are characterized by long durations. For example, a part or a component will generally have a different failure rate in power-off and power-on conditions [2,5,11].

There are various computer-related reliability measures, which may be grouped under the following classifications [2,5,7]:

- *Classification 1*: This classification contains those measures that are considered quite suitable for networks and configurations such as standby, hybrid, and massively redundant systems. The measures are reliability, mean time to failure (MTTF), availability, and mission time. It is to be noted that these measures may not be sufficient to evaluate degrading systems gracefully.
- *Classification 2*: This classification contains the following measures for gracefully handling degrading systems:
  - *Computation reliability*: This is the probability that the system will, without error, execute a task of length (say, y) that started at time t.
  - *Computation availability*: This is the system's expected computation capacity at a stated time t.
  - *Mean computation before failure*: This is the expected amount of computation available on the system prior to failure.
  - *Computation threshold*: This is the time at which a certain value of computation reliability is reached for a task whose length is, say, z.
  - *Capacity threshold*: This is the time at which a certain value of computation availability is reached.

7.4   Fault masking

In the area of fault-tolerant computing, the term *fault masking* is used in the sense that a system with redundancy can tolerate a number of failures before it fails altogether. Thus, the implication of the term *masking* is that some kind of problem has occurred somewhere within the framework of a digital system, but because of the design, the problem does not affect the system's overall operation. Probably the best-known fault-masking approach/method is modular redundancy.

7.4.1   Triple modular redundancy (TMR)

With triple modular redundancy (TMR), three identical modules/units carry out the same task simultaneously and a voter compares the output of the modules or units' and sides with the majority. The TMR system

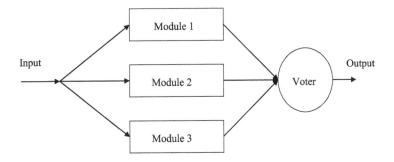

Figure 7.1   Diagram of a TMR system with voter block.

malfunctions when at least two modules/units fail or the voter fails. It simply means that the TMR system can tolerate a single module/unit failure.

The block diagram of the TMR system with the voter is shown in Figure 7.1 [2,5,7,12].

For independent voter and module units, the reliability of the TMR system with the voter is expressed by the following equation [2,5,12]:

$$R_{ts} = (3R_{mu}^2 - 2R_{mu}^3)R_{vo}, \tag{7.1}$$

where R_{ts} is the TMR system with voter reliability, R_{mu} is the module/unit reliability, and R_{vo} is the voter unit reliability.

For the constant failure rates of the TMR system modules and units and the voter unit, and with the aid of the material presented in Chapter 3 and Equation 7.1, we obtain

$$\begin{aligned} R_{ts}(t) &= (3e^{-2\lambda_{mu}t} - 2e^{-3\lambda_{mu}t})e^{-\lambda_{vo}t} \\ &= 3e^{-(2\lambda_{mu}+\lambda_{vo})t} - 2e^{-(3\lambda_{mu}+\lambda_{vo})t}, \end{aligned} \tag{7.2}$$

where λ_{mu} is the module/unit constant failure rate, λ_{vo} is the voter unit constant failure rate, and $R_{ts}(t)$ is the TMR system with voter reliability at time t.

By integrating Equation 7.2 over the time interval $[0, \infty]$, we get the following equation for the TMR system with voter MTTF [2,5,12]:

$$\begin{aligned} \text{MTTF}_{ts} &= \int_0^\infty \left[3e^{-(2\lambda_{mu}+\lambda_{vo})t} - 2e^{-(3\lambda_{mu}+\lambda_{vo})t} \right] dt \\ &= \frac{3}{(2\lambda_{mu} + \lambda_{vo})} - \frac{2}{(3\lambda_{mu} + \lambda_{vo})}, \end{aligned} \tag{7.3}$$

where MTTF_{ts} is the TMR system with voter MTTF.

EXAMPLE 7.1

Assume that the constant failure rate of a module or unit of a TMR system with a voter is 0.004 failures/h and the voter's constant failure rate is 0.001 failures/h.

Calculate the TMR system's reliability for a 200-h mission and MTTF.

By placing the given data values into Equation 7.2, we obtain

$$R_{ts}(t) = 3e^{-[2(0.004) + 0.001](200)} - 2e^{-[3(0.004) + 0.001](200)}$$
$$= 0.3473.$$

Similarly, by putting the given data values into Equation 7.3, we obtain

$$= \frac{3}{[2(0.004) + 0.001]} - \frac{2}{[3(0.004) + 0.001]}$$
$$= 179.48 \text{ h}$$

Thus, the TMR system's reliability and MTTF are 0.3473 and 179.48 h, respectively.

7.4.2　N-modular redundancy (NMR)

N-modular redundancy (NMR) is the general form of the TMR that contains N identical modules or units instead of just 3 of them. The number N is an odd number expressed by $N = 2k + 1$. The NMR system will be successful (or at least functional) if at least $(k + 1)$ modules/units function normally. As the voter unit acts in series with an N-module system, the entire system fails whenever the voter unit malfunctions.

For independent modules and voter units, the reliability of the NMR system with a voter is expressed by the following equation [2,5,12,13]:

$$R_{Nts} = \left[\sum_{j=0}^{m} \binom{N}{j} R_{mu}^{N-1}(1 - R_{mu})^j \right] R_{vo}, \tag{7.4}$$

where

$$\binom{N}{j} = \frac{N!}{(N-J)! \, j!},$$

R_{Nts} is the NMR system with voter reliability, R_{mu} is the module/unit reliability, and R_{vo} is the voter unit reliability.

Note that time-dependent reliability analysis of an NMR system can be conducted in a similar way as the TMR system time-dependent reliability analysis presented earlier in this chapter. Information on additional redundancy schemes is available in Reference 11.

7.5   *Internet facts, statistics, and failure examples, and observations related to Internet reliability*

Several facts, statistics, and failure examples concerning Internet are as follows:

- During the period 2006–2011, developing countries around the globe increased their share of the world's total number of Internet users from 44% to 62% [5,14].
- In 2011, more than 2.1 billion people in the world were using the Internet, and approximately 45% of them were below the age of 25 years [14].
- In 2000, the Internet carried 51% of the information flowing through two-way telecommunication, and by 2007, over 97% of all telecommunicated information was carried over the Internet [15].
- In 2011 alone, there were 52,658 Internet-related incidents and failures [4,5].
- In 2000, in the United States, the Internet-related economy generated around $830 billion in revenue [4,5].
- *Example*: On August 24, 1998, a misconfigured main Internet database server wrongly referred all queries for Internet systems and machines with names ending in ".net" to the incorrect secondary database server. In turn, due to this problem, most of the connections to ".net" Internet web servers and other end stations malfunctioned for many hours [4,5].
- *Example*: On November 8, 1998, a malformed routing control message because of a software fault triggered an interoperability problem between a number of core Internet backbone routers produced by different vendors. In turn, this resulted in a widespread loss of network connectivity, in addition to an increment in packet loss and latency [3,5].
- *Example*: On April 25, 1997, a misconfigured router of a Virginia service provider injected the wrong map into the global Internet and, in turn, the Internet providers who accepted this map automatically diverted their traffic to the Virginia provider [3,16]. This resulted in network congestion, instability, and overload of Internet router table memory that ultimately shut down many of the main Internet backbones for approximately 2 h [3,5,16].

A study reported the following observations related to Internet reliability [2,17]:

- *Observation I:* Most of the Internet backbone paths' MTTF and mean time to repair (MTTR) are around 25 days or less and 20 minutes or less, respectively.
- *Observation II:* The Internet backbone structure's MTTF and availability are quite significantly less than of the Public Switched Telephone Network (PSTN).
- *Observation III:* Most interprovider path malfunctions occur from congestion collapse.
- *Observation IV:* In the Internet backbone infrastructure, there are only a small number of network paths that disproportionately contribute to long-term outages and backbone unavailability.

7.6 Classifications of Internet-related outages and an approach for automating fault detection in Internet services

Past experiences over the years indicate that there are many types of Internet-related outages. A case study of Internet-related outages performed over a 1-year period categorized the outages under 12 classifications (along with their occurrence percentages in parentheses), as shown in Figure 7.2 [2,5,17].

Past experience indicates that many Internet services (e.g., e-commerce and search engines) suffer faults, and a quick detection of these faults could be an important factor to improve system availability. Thus, for this very purpose, an approach referred to as the *pinpoint method* is considered very helpful. This method combines the low-level monitors' easy deployability with the higher-level monitors' ability to detect application-level faults [2,5,18]. With respect to the system under observation and its workload, the method is based upon the following assumptions [2,5,18]:

- The software is made up of a number of interconnected modules with clearly defined narrow interfaces, which could be software subsystems, objects, or simply physical mode boundaries.
- An interaction with the system is short-lived, and its processing could be decomposed as a path or, more specifically, a tree of the names of parts or elements that participate in the servicing of that request.
- There are a considerably higher number of basically independent requests from different users.

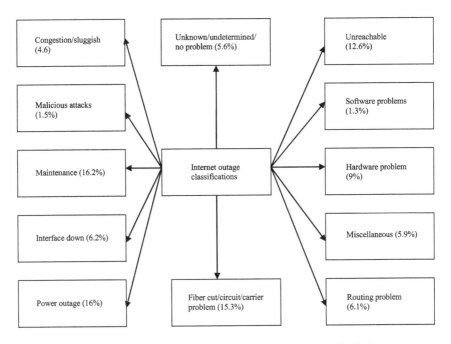

Figure 7.2   Classifications of Inter-related outages (along with their occurrence percentages, in parentheses).

The pinpoint method is a three-stage process, as described next [2,5,18]:

1. *Stage I: Observing the system*—This is concerned with capturing the runtime path of each and every request handled/served by the system and then, from these very paths, extracting two particular low-level behaviours that are quite likely to reflect high-level functionality (i.e., interactions of components/parts as well as path shapes).
2. *Stage II: Learning the patterns in system behaviour*—This is concerned with constructing a reference model that clearly represents the normal behaviour of an application with respect to part/component interactions and path shapes. The model is constructed under the assumption that most of the time, the system operates normally.
3. *Stage III: Detecting anomalies in system behaviours*—This is concerned with analyzing the system ongoing behaviours and detecting anomalies in regard to the reference model.

All in all, additional information on this method is available in Reference 18.

7.7   *Mathematical models for performing Internet reliability and availability analysis*

Many mathematical models can be used for performing various types of availability and reliability analysis concerning the availability and reliability of Internet services [2,3,5,12,19–22]. Two such models are presented next.

7.7.1   *Model I*

This mathematical model is concerned with evaluating the availability and reliability of an Internet server system. The model assumes that the Internet server system can be in either an operating or a failed state. In addition, the model assumes that all its failures and outages occur independently, the restored or repaired server system is as good as new, and its outage/failure and restoration/repair rates are constant.

The Internet server system state space diagram is shown in Figure 7.3, and the numerals in the boxes denote the system states.

The following symbols are associated with this mathematical model:

j is the jth Internet server system state shown in Figure 7.3 for $j = 0$ (meaning that the Internet server system is operating normally), and $j = 1$ (meaning that the Internet server system failed).
λ_{is} is the Internet server system's constant failure/outage rate.
μ_{is} is the Internet server system's constant repair/restoration rate.
$P_j(t)$ is the probability that the Internet server system is in state j at time t, for $j = 0, 1$.

By using the Markov method presented in Chapter 4, we write the following differential equations for the state space diagram shown in Figure 7.3 [2,5,12]:

$$\frac{dP_0(t)}{dt} + \lambda_{is}P_0(t) = \mu_{is}P_1(t) \tag{7.5}$$

$$\frac{dP_1(t)}{dt} + \mu_{is}P_1(t) = \lambda_{is}P_0(t). \tag{7.6}$$

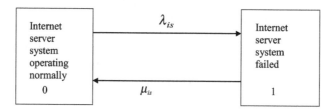

Figure 7.3   State space diagram of an Internet server system.

At time $t = 0$, $P_0(0) = 1$, and $P_1(0) = 0$.
By solving Equations 7.5 and 7.6, we obtain the following equations:

$$P_0(t) = AV_{is}(t) = \frac{\mu_{is}}{(\lambda_{is} + \mu_{is})} + \frac{\lambda_{is}}{(\lambda_{is} + \mu_{is})} e^{-(\lambda_{is} + \mu_{is})t} \tag{7.7}$$

$$P_1(t) = UA_{is}(t) = \frac{\lambda_{is}}{(\lambda_{is} + \mu_{is})} - \frac{\lambda_{is}}{(\lambda_{is} + \mu_{is})} e^{-(\lambda_{is} + \mu_{is})t}, \tag{7.8}$$

where $AV_{is}(t)$ is the Internet server system availability at time t and $UA_{is}(t)$ is the Internet server system unavailability at time t.
As time t becomes very large, Equations 7.7 and 7.8 reduce to

$$AV_{is} = \lim_{t \to \infty} AV_{is}(t) = \frac{\mu_{is}}{\lambda_{is} + \mu_{is}} \tag{7.9}$$

$$UA_{is} = \lim_{t \to \infty} UA_{is}(t) = \frac{\lambda_{is}}{\lambda_{is} + \mu_{is}}, \tag{7.10}$$

where AV_{is} is the Internet server system steady state availability and UA_{is} is the Internet server system steady state unavailability.
For $\mu_{is} = 0$, Equation 7.7 becomes

$$R_{is}(t) = e^{-\lambda_{is}t}, \tag{7.11}$$

where $R_{is}(t)$ is the Internet server system reliability at time t.
Thus, the Internet server system MTTF is given by [12]

$$\begin{aligned}
\mathrm{MTTF}_{is} &= \int_0^\infty R_{is}(t)dt \\
&= \int_0^\infty e^{-\lambda_{is}t}dt \\
&= \frac{1}{\lambda_{is}},
\end{aligned} \tag{7.12}$$

where MTTF_{is} is the Internet server system MTTF.

EXAMPLE 7.2

Assume that the constant failure and repair rates of an Internet server system are 0.008 failures/h and 0.09 repairs/h, respectively. Calculate the Internet server system unavailability for a 20-h mission.

By substituting the specified data values into Equation 7.8, we obtain

$$UA_{is}(20) = \frac{0.008}{(0.008 + 0.09)} - \frac{0.008}{(0.008 + 0.09)} e^{-(0.008+0.09)(20)}$$

$$= 0.0701.$$

Thus, the Internet server system unavailability for the specified mission time is 0.0701.

7.7.2 Model II

This mathematical model is concerned with evaluating the availability of an Internet working (router) system composed of two independent and identical switches. The model assumes that the switches form a standby-type network and that the system fails when both switches fail. Furthermore, the switch failure and repair and restoration rates are constant. The system state space diagram is shown in Figure 7.4, and the numerals in boxes denote the system states.

The following symbols were used for developing equations for the model:

j is the jth system state shown in Figure 7.4, for $j = 0$ [system operating normally (i.e., two switches functional: one operating, the other on standby)] and $j = 1$ (one switch operating, the other failed), $j = 2$ [system failed (both switches failed)].

$P_j(t)$ is the probability that the Internet working (router) system is in state j at time t, for $j = 0, 1, 2$.

λ_s is the switch constant failure rate.

γ_s is the switch constant restoration and repair rate.

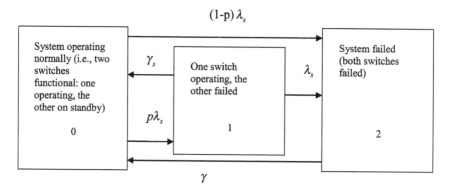

Figure 7.4   System state space diagram.

γ is the constant restoration and repair rate from system state 2 to state 0.

p is the probability of failure detection and a successful switchover from switch failure.

By using the Markov method presented in Chapter 4, we write the following differential equations for the diagram shown in Figure 7.4 [2,5,12,23]:

$$\frac{dP_0(t)}{dt} + [p\lambda_s + (1-p)\lambda_s]P_0(t) = \gamma_s P_1(t) + \gamma P_2(t) \tag{7.13}$$

$$\frac{dP_1(t)}{dt} + (\lambda_s + \gamma_s)P_1(t) = p\lambda_s P_0(t) \tag{7.14}$$

$$\frac{dP_2(t)}{dt} + \gamma P_2(t) = \lambda_s P_1(t) + (1-p)\lambda_s P_0(t). \tag{7.15}$$

At time $t = 0$, $P_0(0) = 1$ and $P_1(0) = P_2(0) = 0$.

The following steady-state probability solutions are obtained by setting derivatives equal to zero in Equations 7.13 through 7.15 and using the relationship $\sum_{j=0}^{2} P_j = 1$:

$$P_0 = \frac{\gamma(\gamma_s + \lambda_s)}{C} \tag{7.16}$$

where

$$C = \gamma(\gamma_s + p\lambda_s + \lambda_s) + (1-p)\lambda_s(\gamma_s + \lambda_s) + p\lambda_s^2 \tag{7.17}$$

$$P_1 = \frac{p\lambda_s\gamma}{C} \tag{7.18}$$

$$P_2 = \frac{[p\lambda_s^2 + (1-p)\lambda_s(\gamma_s + \lambda_s)]}{C}, \tag{7.19}$$

where P_j is the steady-state probability that the Internet working (router) system is in state j for $j = 0, 1, 2$.

The Internet working (router) system steady-state availability is given by

$$AV_{iss} = P_0 + P_1$$
$$= \frac{[\gamma(\gamma_s + \lambda_s) + p\lambda_s\gamma]}{C}, \tag{7.20}$$

where AV_{iss} is the Internet working (router) system steady-state availability.

7.8   Methods for preventing programmers from introducing bugs during the software-writing process

There are many methods for preventing programmers from introducing bugs during the software-writing process. Four of these methods are as follows [5,24,25]:

1. *Method I: Development methodologies*—There are several methodologies and schemes for managing programmer-related activity so that fewer bugs are generated. Most of these fall under the software-engineering discipline (which examines software design–related issues as well). For example, formal program specifications are employed for stating the exact behavior of all involved programs so that design-related bugs can be eradicated.
2. *Method II: Programming language support*—In this case, programming languages include features such as modular programming, restricted name spaces, and static type systems to help programmers prevent bugs.
3. *Method III: Code analysis*—In this case, tools for code analysis help software developers by inspecting the program text beyond the compiler's capability for highlighting potential problems or difficulties. Although the problem of finding all types of programming-related errors in a given specification is generally not solvable, these tools exploit the fact that programmers tend to make the same kinds of errors during software writing.
4. *Method IV: Programming style*—In this case, various innovations in programming style and defensive programming are designed for making the occurrence of software bugs less or more likely to spot.

Additional information on these four methods is available in References 24 and 25.

7.9   Metrics related to software errors

Over the years, in order to measure the extent of errors, faults, and failures in system artifacts during the life-cycle phases, many metrics have been developed. Three of these, considered quite useful for dealing with software errors, are presented next [5,26–28].

7.9.1   Metric 1: Fault density

Metric 1 calculates the ratio of faults per line of code and is defined by [5,26–28]

$$F_d = \frac{n_u}{\theta},\qquad(7.21)$$

where F_d is the fault density, n_u is the number of unique faults found, and θ is the lines of executable source code (in thousands), including data declarations.

The metric can be utilized during all life-cycle phases [28]. Two variations to this metric can be made in order to enhance its usefulness for mission-critical systems [26]. First, to minimize the problems encountered with lines of code (LOCs), an equivalent calculation can be conducted by inserting function points (FPs) for θ. Second, if the fault's severity is distinctive, the calculation can be expanded or enlarged for determining fault density by severity. Finally, note that Reference 28 gives this metric an experience code of 2, indicating that it has received a reasonable level of operational validity in the industrial sector.

Additional information on this metric is available in Reference 26.

7.9.2   Metric 2: Defect density

Metric 2 calculates the ratio of defects per lines of design or lines of code and is defined by

$$\alpha = \frac{\left(\sum_{i=1}^{m} UD_i\right)}{M(\text{or } N)},$$

(7.22)

where α is the defect density, m is the number of inspections (or life-cycle phases) to date, UD_i is the number of unique defects found during the ith inspection process of a life-cycle phase, M is the total number of executable source code statements plus data declarations during the implementation phase and beyond, and N is the total number of source lines of design-related statements (in thousands) during the design phase.

This metric can be utilized during all eight life-cycle phases [28]. Furthermore, Reference 28 gives the metric an experience code of 3, indicating that it has received a high level of validity in the industrial sector.

Additional information on this metric is available in References 26–28.

7.9.3   Metric 3: Defect indices

Metric 3 calculates a relative index of a software program's correctness throughout the various life-cycle phases and is expressed by [5,26–28]

$$DI = \frac{\sum i Y_i}{A_i},$$

(7.23)

where DI is the defect index; A_i is the product size at ith phase, which can be measured in M, N, or function points; and

$$Y_i = [\alpha_1 \lambda_i / \mu_i] + [\alpha_2 \beta_i / \mu_i] + [\alpha_3 \gamma_i / \mu_i],$$

(7.24)

where λ_i is the number of serious defects found during the ith phase; μ_i is the total number of defects detected during the ith phase; β_i is the number of medium defects found during the ith phase; γ_i is the number of trivial defects found during the ith phase; α_1 is the weighting factor for serious defects (default is 10); α_2 is the weighting factor for medium defects (default is 3); and α_3 is the weighting factor for trivial defects (default is 1).

Note that at the end of each of the eight phases, the value of Y_i is calculated and is weighted by phase, such that $i = 1, 2, 3, \ldots, 8$. The metric from Equation 7.23 can be utilized in all life-cycle phases [28]. Furthermore References 2 and 8 gives this metric an experience code of 1, indicating that in the industrial sector, it has received a low level of operational validity.

Additional information on this metric is available in References 26–28.

PROBLEMS

1. Discuss at least six main sources of computer failures.
2. Describe the following terms:
    i.   N-modular redundancy (NMR)
    ii. Triple modular redundancy (TMR)
    iii. Fault masking
3. Assume that the constant failure rate of a module of a TMR system with a voter is 0.002 failures/h and the voter's constant failure rate is 0.004 failures/h. Calculate the TMR system's MTTF and reliability for a 100-h mission.
4. Discuss at least five Internet facts, statistics, and failure examples.
5. Write an essay on computer system and Internet reliability.
6. Describe an approach for automating fault detection in Internet services.
7. Prove Equations 7.7 and 7.8 by using Equations 7.5 and 7.6.
8. Prove that the sum of Equations 7.16, 7.18, and 7.19 is equal to unity.
9. Describe at least four methods to prevent programmers from introducing bugs during the software-writing process.
10. Describe the following two software error–related metrics:
    i. Fault density
    ii. Defect density

References

1. Hafner, K., and Lyon, M., *Where Wizards Stay Up Late: The Origin of the Internet*, Simon and Schuster, New York, 1996.
2. Dhillon, B. S., *Engineering Systems Reliability, Safety, and Maintenance: An Integrated Approach*, CRC Press, Boca Raton, FL, 2017.
3. Dhillon, B. S., *Applied Reliability and Quality: Fundamentals, Methods, and Procedures*, Springer-Verlag, London, 2007.

4. Goseva-Popstojanova, K., Mazidar, S., and Singh, A. D., Empirical Study of Session-Based Workload and Reliability for Weber Servers, *Proceedings of the 15th International Symposium on Software Reliability Engineering*, 2004, pp. 403–414.
5. Dhillon, B. S., *Computer System Reliability: Safety and Usability*, CRC Press, Boca Raton, FL, 2013.
6. Goldberg, J., A Survey of the Design and Analysis of Fault-Tolerant Computers, in *Reliability and Fault Tree Analysis*, edited by R. E. Barlow, J. B. Fussell, and N. D. Singpurwalla, Society for Industrial and Applied Mathematics, Philadelphia, 1975, pp. 667–685.
7. Dhillon, B. S., *Reliability in Computer System Design*, Ablex Publishing, Noorwood, NJ, 1987.
8. Yourdon, E., The Causes of Computer Failures: Part II, *Modern Data*, Vol. 5, 1972, pp. 36–40.
9. Yourdon, E., The Causes of Computer Failures: Part III, *Modern Data*, Vol. 5, 1972, pp. 50–56.
10. Avizienis, A., Fault-Tolerant Computing: Progress, Problems, and Prospectus, *Proceedings of the International Federation of Information Processing Congress*, 1977, pp. 405–420.
11. Nerber, P. O., Power-Off Time Impact on Reliability Estimates, *IEEE International Convention Record*, Part 10, March 1965, pp. 1–8.
12. Dhillon, B. S., *Design Reliability: Fundamentals and Applications*, CRC Press, Boca Raton, FL, 1999.
13. Shooman, M. L., *Reliability of Computer Systems and Networks: Fault Tolerance, Analysis, and Design*, John Wiley and Sons, New York, 2002.
14. ICT Fact and Figures, *ICT Data and Statistics Division*, Telecommunication Development Bureau, International Telecommunication Union, Geneva, Switzerland, 2011.
15. Hilbert, M., and Lopez, P., The World's Technological Capacity to Store, Communicate, and Compute Information, *Science*, 332 (6025), April 2011, pp. 60–65.
16. Barrett, R., Haar, S., and Whitestone, R., Routing Snafu Causes Internet Outage, *Interactive Week*, April 25 1997, p. 9.
17. Lapovitz, C., Ahuja, A., and Jahamian, F., Experimental Study of the Internet Stability and Wide-Area Backbone Failures, *Proceedings of the 29th Annual International Symposium on Fault-Tolerant Computing*, 1999, pp. 278–285.
18. Kiciman, E., and Fox, A., Detecting Application-Level Failures in Component-Based Internet Services, *IEEE Transactions on Neural Networks*, Vol. 16, No. 5, 2005, pp. 1027–1041.
19. Imaizumi, M., Kimura, M., and Yasui, K., Optimal Monitoring Policy for Server System with Illegal Access, *Proceedings of the ISSAT International Conference on Reliability and Quality in Design*, 2005, pp. 155–159.
20. Aida, M., and Abe, T., Stochastic Model of Internet Access Patterns, *IEICE Transactions on Communications*, Vol. E 84-B(8), 2001, pp. 2142–2150.
21. Chan, C. K., and Tortorella, M., Spare-Inventory Sizing from End-to-End Service Availability, *Proceedings of the Annual Reliability and Maintainability Symposium*, 2001, pp. 98–102.

22. Hecht, M., Reliability/Availability Modeling and Prediction of e-Commerce and Other Internet Information Systems, *Proceedings of the Annual Reliability and Maintainability Symposium*, 2001, pp. 176–182.

23. Dhillon, B. S., and Kirmizi, F., Probabilistic Safety Analysis of Maintainable Systems, *Journal of Quality in Maintenance Engineering*, Vol. 9, No.3, 2003, pp. 303–320.

24. McDonald, M., Musson, R., and Smith, R., *The Practical Guide to Defect Prevention*, Microsoft Press, Seattle, 2007.

25. Huizinga, D., and Kolawa, A., *Automated Defect Prevention: Best Practices in Software Management*, Wiley-IEEE Computer Society Press, New York, 2007.

26. Herrman, D. S., *Software Safety and Reliability*, IEEE Computer Society Press, Los Alamitos, California, 1999.

27. ANSI / IEEE STD 982.2-1989, *IEEE Guide for the Use of IEEE Standard Dictionary of Measures to Produce Reliable Software*, IEEE, New York, 1989.

28. ANSI / IEEE STD 982.1-1989, *IEEE Standard Dictionary of Measures to Produce Reliable Software*, IEEE, New York, 1989.

chapter eight

Power system and mining equipment reliability

8.1 Introduction

The primary function of a modern electric power system is supplying its customers cost-effective electrical energy with a high degree of reliability. During planning, design, control, operation, and maintenance of an electric power system, consideration of the two important aspects of quality and continuity of supply, along with other pertinent factors, is generally referred to as *reliability assessment*.

In the context of an electric power system, *reliability* may simply be expressed as regarding the ability of the system for providing a satisfactory amount of electrical power [1]. The history of power system reliability goes back to the early 1930s, when probability concepts were applied to electric power–associated problems [2–5].

Each year, billions of dollars are spent to produce various types of equipment for use by the mining industrial sector around the globe, and this expenditure is increasing significantly [6]. Furthermore, as mining equipment is becoming more complex and sophisticated, its cost is increasing quite rapidly. In order to meet production targets, mining companies are increasingly demanding better equipment reliability.

This chapter presents various important aspects of power system and mining equipment reliability.

8.2 Power system reliability–related terms and definitions

There are many power system reliability–related terms and definitions. Some of the most common ones are presented here [5,7–9]:

- *Power system reliability*: This is the degree to which the performance of the elements in a bulk system results in electrical energy being delivered to customers within the framework of stated standards and in the amount needed.
- *Forced outage*: This is when a piece of equipment or unit has to be taken out of service because of damage or component failure.

- *Forced outage rate*: This statistic (applying to equipment) is the total number of forced outage hours times 100 over the total number of service hours, plus the total number of forced outage hours.
- *Service hours*: These are the total number of operation hours of a piece of equipment or unit.
- *Mean time to forced outage*: This is analogous to mean time to failure (MTTF) and is given by the total number of service hours over the total number of forced outages.
- *Forced outage hours*: These are the total number of hours that a piece of equipment/unit spends in the forced outage condition.
- *Mean forced outage duration*: This is analogous to mean time to repair (MTTR) and is given by the total number of forced outage hours over the total number of forced outages.

8.3   *Service performance–related indexes*

In the area of electric power system, there are many service performance–related indexes. Four of these indexes are described in the following subsections [1,5,10].

8.3.1   *Index I*

Index I is referred to as the *system average interruption frequency index (SAIFI)* and is expressed by

$$\text{SAIFI} = \frac{\alpha}{\theta},\tag{8.1}$$

where θ is the total number of customers and α is the total number of customer interruptions per year.

8.3.2   *Index II*

Index II is referred to as the *system average interruption duration index (SAIDI)* and is defined by

$$\text{SAIDI} = \frac{\text{SCID}}{\theta},\tag{8.2}$$

where SCID is the sum of customer interruption durations per year.

8.3.3   *Index III*

Index III is referred to as the *average service availability index (ASAI)* and is defined by

$$\text{ASAI} = \frac{\gamma}{\beta},\tag{8.3}$$

where β is customer hours demanded (with these hours given by the 12-month average number of customers serviced times 8,760 h), and γ is the customer hours of available service.

8.3.4   Index IV

Index IV is referred to as the *customer average interruption frequency index* (CAIFI) and is expressed by

$$\mathrm{CAIFI} = \frac{\alpha}{C_{tn}}, \tag{8.4}$$

where C_{tn} is the total number of customers affected. Note that the customers affected should be counted only once, regardless of the number of interruptions throughout the year that they may have experienced.

8.4   Loss-of-load probability (LOLP)

LOLP is the single most important metric used to estimate overall power system reliability, and it may simply be described as a projected value of how much time, in the long run, the load on a given power system is expected to be greater than the capacity of the generating resources [5,7]. Various probabilistic methods are used for calculating LOLP.

In the setup of an LOLP criterion, it is assumed that an electric power system is strong enough to have a low LOLP and can probably withstand most of the foreseeable peak loads, contingencies, and outages. Thus, a utility is expected to arrange for resources (i.e., generation, load management, purchases, etc.) in such a manner that the resulting LOLP will be at or less than the acceptable level.

In the power generation sector, the common practice generally is to plan that the power system will achieve an LOLP of 0.1 days per year or less. All in all, some of the difficulties experienced with this use of LOLP are presented here [5,7]:

- LOLP itself does not state the duration or magnitude of the shortage of electricity.
- LOLP does not take into consideration the factor of additional emergency support that one region or control area may receive from another, or other emergency-related measures or actions that control area operators can take to maintain system reliability.
- Different LOLP estimation techniques can lead to different indexes for the exact same electric power system.
- Major loss-of-load incidents take place because of contingencies that are not modeled appropriately by the traditional LOLP calculation.

8.5 Mathematical models for performing availability analysis of a single generator unit

There are a number of mathematical models that can be used for performing availability analysis of a single generator unit. Two such models are presented in the next subsections.

8.5.1 Model I

Model I represents a generator unit that can either be in operating state or failed state, and it is repaired from the failed state. The state space diagram of the generator unit is shown in Figure 8.1. The single letters in boxes denote the generator unit states.

The model is subjected to the following assumptions:

- The generator unit failure and repair rates are constant.
- The repaired generator unit is as good as new.
- The generator unit failures occur independently.

The following symbols are associated with the diagram shown in Figure 8.1 and its associated equations:

$P_j(t)$ is the probability that the generator unit is in state j at time t; for $j = 0$ (operating normally), $j = f$ (failed).
λ is the generator unit constant failure rate.
μ is the generator unit constant repair rate.

With the aid of the Markov method described in Chapter 4, we write the following equations for the diagram in Figure 8.1 [5,11]:

$$\frac{dP_o(t)}{dt} + \lambda P_o(t) - \mu P_f(t) = 0 \tag{8.5}$$

$$\frac{dP_f(t)}{dt} + \mu P_f(t) - \lambda P_o(t) = 0. \tag{8.6}$$

At time $t = 0$, $P_o(0) = 1$, and $P_f(0) = 0$.

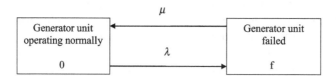

Figure 8.1 Generator unit state space diagram.

By solving Equations 8.5 and 8.6, we obtain

$$P_o(t) = \frac{\mu}{\lambda + \mu} + \frac{\lambda}{\lambda + \mu} e^{-(\lambda+\mu)t} \tag{8.7}$$

$$P_f(t) = \frac{\lambda}{\lambda + \mu} - \frac{\mu}{\lambda + \mu} e^{-(\lambda+\mu)t}. \tag{8.8}$$

The generator unit availability and unavailability are given by

$$AV_{gu}(t) = P_o(t) = \frac{\mu}{\lambda + \mu} + \frac{\lambda}{\lambda + \mu} e^{-(\lambda+\mu)t} \tag{8.9}$$

and

$$UA_{gu}(t) = P_f(t) = \frac{\lambda}{\lambda + \mu} - \frac{\mu}{\lambda + \mu} e^{-(\lambda+\mu)t}, \tag{8.10}$$

where $AV_{gu}(t)$ is the generator unit availability at time t and $UA_{gu}(t)$ is the generator unit unavailability at time t.

For a very large t, Equations 8.9 and 8.10 reduce to

$$AV_{gu} = \frac{\mu}{\lambda + \mu} \tag{8.11}$$

and

$$UA_{gu} = \frac{\lambda}{\lambda + \mu}, \tag{8.12}$$

where AV_{gu} is the generator unit's steady-state availability and UA_{gu} is the generator unit's steady-state unavailability.

Because $\lambda = (1/\text{MTTF})$ and $\mu = (1/\text{MTTR})$, Equations 8.11 and 8.12 become

$$AV_{gu} = \frac{\text{MTTF}}{\text{MTTR} + \text{MTTF}} \tag{8.13}$$

and

$$UA_{gu} = \frac{\text{MTTR}}{\text{MTTR} + \text{MTTF}} \tag{8.14}$$

EXAMPLE 8.1

Assume that the constant failure and repair rates of a generator unit are $\lambda = 0.0004$ failures/h and $\mu = 0.0008$ repairs/h, respectively. Calculate the generator unit's steady-state unavailability.

By inserting the given data values into Equation 8.12, we obtain

$$UA_{gu} = \frac{0.0004}{0.0004 + 0.0008} = 0.3333.$$

Thus, the generator unit's steady-state unavailability is 0.3333.

8.5.2   Model II

Model II represents a generator unit that can either be in operating state or a failed state or taken out of service for preventive maintenance, and its state space diagram is shown in Figure 8.2. The single letters in circles and box denote the generator unit states.

The model is subject to the following assumptions:

- The generator unit failures occur independently.
- The generator unit failure, repair, preventive maintenance down, and preventive maintenance performance rates are constant.
- After preventive maintenance and repair, the generator unit is as good as new.

The following symbols are associated with the diagram shown in Figure 8.2 and its associated equations:

$P_j(t)$ is the probability that the generator unit is in state j at time t; for $j = $ o (operating normally) and $j = p$ (down for preventive maintenance), $j = f$ (failed).
λ_f is the generator unit's failure rate.
λ_m is the generator unit's (down for) preventive maintenance rate.
μ_s is the generator unit's repair rate.
μ_m is the generator unit's preventive maintenance performance rate.

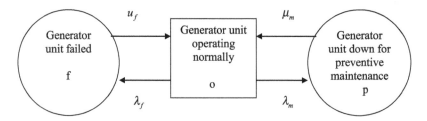

Figure 8.2   Generator unit state space diagram.

With the aid of the Markov method described in Chapter 4, we write the following equations for the diagram in Figure 8.2 [5,11]:

$$\frac{dP_o(t)}{dt} + (\lambda_f + \lambda_m)P_o(t) - \mu_f P_f(t) - \mu_m P_p(t) = 0 \tag{8.15}$$

$$\frac{dP_p(t)}{dt} + \mu_p P_p(t) - \lambda_m P_o(t) = 0 \tag{8.16}$$

$$\frac{dP_f(t)}{dt} + \mu_f P_f(t) - \lambda_f P_o(t) = 0. \tag{8.17}$$

At time $t = 0$, $P_o(0) = 1$, $P_p(0) = 0$, and $P_f(0) = 0$.
By solving Equations 8.15 through 8.17, we obtain

$$P_o(t) = \frac{\mu_m \mu_f}{n_1 n_2} + \left[\frac{(n_1 + \mu_m)(n_1 + \mu_f)}{n_1(n_1 - n_2)}\right]e^{n_1 t} - \left[\frac{(n_2 + \mu_m)(n_2 + \mu_f)}{n_2(n_1 - n_2)}\right]e^{n_2 t} \tag{8.18}$$

$$P_p(t) = \frac{\lambda_m \mu_f}{n_1 n_2} + \left[\frac{(\lambda_m n_1 + \lambda_m \mu_s)}{n_1(n_1 - n_2)}\right]e^{n_1 t} - \left[\frac{(\mu_f + n_2)\lambda_m}{n_2(n_1 - n_2)}\right]e^{n_2 t} \tag{8.19}$$

$$P_f(t) = \frac{\lambda_f \mu_m}{n_1 n_2} + \left[\frac{(\lambda_f n_1 + \lambda_f \mu_m)}{n_1(n_1 - n_2)}\right]e^{n_1 t} - \left[\frac{(\mu_m + n_2)\lambda_f}{n_2(n_1 - n_2)}\right]e^{n_2 t}, \tag{8.20}$$

where

$$n_1 n_2 = \mu_m \mu_f + \lambda_m \mu_f + \lambda_f \mu_m \tag{8.21}$$

$$n_1 + n_2 = -(\mu_m + \mu_f + \lambda_m + \lambda_f). \tag{8.22}$$

The generator unit availability is given by

$$GUAV(t) = P_o(t)$$
$$= \frac{\mu_m \mu_f}{n_1 n_2} + \left[\frac{(n_1 + \mu_m)(n_1 + \mu_f)}{n_1(n_1 - n_2)}\right]e^{n_1 t}$$
$$- \left[\frac{(n_2 + \mu_m)(n_2 + \mu_f)}{n_2(n_1 - n_2)}\right]e^{n_2 t}, \tag{8.23}$$

where $GUAV(t)$ is the generator unit availability at time t.
Note that this availability expression is valid if and only if n_1 and n_2 are negative. Thus, for large t, Equation 8.23 reduces to

$$GUAV = \lim_{t \to \infty} GUAV(t) = \frac{\mu_n \mu_f}{n_1 n_2}, \tag{8.24}$$

where GUAV is the generator unit steady-state availability.

EXAMPLE 8.2

Assume that for a generator unit, we have the following given data values:

$\lambda_f = 0.0001$ failures/h
$\mu_f = 0.0004$ repairs/h
$\lambda_m = 0.0006$ failures/h
$\mu_m = 0.0008$ repairs/h

Calculate the steady-state availability of the generator unit.
By substituting the given data values into Equation 8.24, we obtain

$$GUAV = \frac{(0.0008)(0.0004)}{(0.0008)(0.0004) + (0.0006)(0.0004) + (0.0001)(0.0008)} = 0.5.$$

Thus, the steady-state availability of the generator unit is 0.5.

8.6 Mathematical models for performing availability analysis of transmission and associated systems

In the area of the power system, various types of equipment and systems are used to transmit electrical energy from one end to another. Two examples of such equipment and systems are transmission lines and transformers.

This section presents two mathematical models that can be used for performing availability analysis of transmission and associated systems [5,8,12].

8.6.1 Model I

Model I represents a system composed of transmission lines and other equipment functioning in outdoor environments (i.e., normal and stormy), and it can fail under both conditions. The state space diagram of the system is shown in Figure 8.3. The numerals in boxes denote system states.

The model is subjected to the following assumptions:

- The system failures occur independently.
- Failure, repair, and environment (weather) fluctuation rates are constant.
- The repaired system is as good as new.

The following symbols are associated with the diagram shown in Figure 8.3 and its associated equations:

$P_j(t)$ is the probability that the system is in state j at time t; for $j = 0$ (operating normally in normal environment), $j = 1$ (failed in normal environment), $j = 2$ (operating normally in stormy environment), $j = 3$ (failed in stormy environment).

θ is the constant transition rate from normal environment to stormy environment.

γ is the constant transition rate from stormy environment to normal environment.

λ_1 is the system constant failure rate in normal environment.

μ_1 is the system constant repair rate in normal environment.

λ_2 is the system constant failure rate in stormy environment.

μ_2 is the system constant repair rate in stormy environment.

With the aid of the Markov method described in Chapter 4, we write the following equations for the diagram in Figure 8.3 [5,12]:

$$\frac{dP_o(t)}{dt} + (\lambda_1 + \theta)P_o(t) - \gamma P_2(t) - \gamma P_2(t) - \mu_1 P_1(t) = 0 \qquad (8.25)$$

$$\frac{dP_1(t)}{dt} + (\mu_1 + \theta)P_1(t) - \gamma P_3(t) - \lambda_1 P_o = 0 \qquad (8.26)$$

$$\frac{dP_2(t)}{dt} + (\lambda_2 + \gamma)P_2(t) - \mu_2 P_3(t) - \theta P_o(t) = 0 \qquad (8.27)$$

$$\frac{dP_3(t)}{dt} + (\gamma + \mu_2)P_3(t) - \lambda_2 P_2(t) - \theta P_1(t) = 0. \qquad (8.28)$$

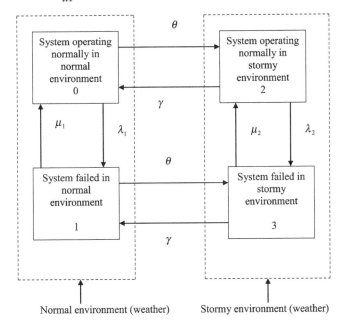

Figure 8.3   System-state space diagram.

At time $t = 0$, $P_o(0) = 1$, $P_1(0) = 0$, $P_2(0) = 0$, and $P_3(0) = 0$.

The following steady-state probability equations are obtained from Equations 8.25 through 8.28 by setting the derivatives with respect to time t equal to zero and using the relationship $\sum_{j=0}^{3} P_j = 1$:

$$P_O = \frac{\gamma A_1}{\theta(A_2 + A_3) + \gamma(A_4 + A_1)}, \tag{8.29}$$

where

$$A_1 = \mu_2\theta + \mu_1 A_5 \tag{8.30}$$

$$A_2 = \mu_1\gamma + \mu_2 A_6 \tag{8.31}$$

$$A_3 = \lambda_1\gamma + \lambda_2 A_6 \tag{8.32}$$

$$A_4 = \lambda_2\theta + \lambda_1 A_5 \tag{8.33}$$

$$A_5 = \lambda_2 + \gamma + \mu_2 \tag{8.34}$$

$$A_6 = \lambda_1 + \theta + \mu_1 \tag{8.35}$$

$$P_1 = \frac{A_4 P_o}{A_1} \tag{8.36}$$

$$P_2 = \frac{\theta P_o A_2}{\gamma A_1} \tag{8.37}$$

$$P_3 = \frac{\theta P_o A_3}{\gamma A_1}. \tag{8.38}$$

where P_o, P_1, P_2, and P_3 are the steady-state probabilities of the system being in states 0, 1, 2, and 3, respectively.

The steady-state availability of the system is given by

$$AV_s = P_o + P_2, \tag{8.39}$$

where AV_s is the system steady-state availability.

8.6.2   Model II

Model II represents a system composed of two nonidentical and redundant transmission lines subject to common-cause failure. A *common-cause failure* may simply be defined as any instance where multiple units fail due to a single cause [13,14]. In transmission lines, a common-cause failure may occur due to various factors, including stormy weather, aircraft crash, and tornado.

The state space diagram of the system is shown in Figure 8.4. The numerals in circles and boxes denote system states.

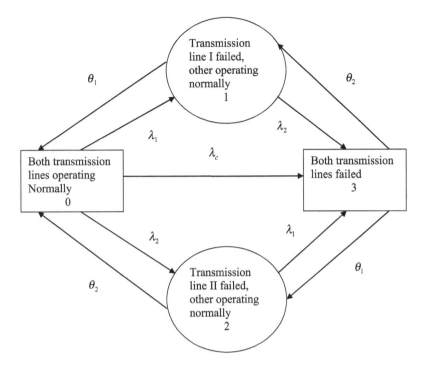

Figure 8.4   System-state space diagram.

The model is subjected to the following assumptions:

- Failure and repair rates of the transmission lines are constant.
- All failures occur independently.
- A repaired transmission line is as good as new.

The following symbols are associated with the diagram shown in Figure 8.4 and its associated equations:

$P_j(t)$ is the probability that the system is in state j at time t; for $j = 0$ (both transmission lines operating normally), $j = 1$ (transmission line I failed, other operating normally), $j = 2$ (transmission line II failed, other operating normally), $j = 3$ (both transmission lines failed).

λ_c is the system constant common-cause failure rate.

λ_1 is the transmission line I constant failure rate.

θ_1 is the transmission line I constant repair rate.

λ_2 is the transmission line II constant failure rate.

θ_2 is the transmission line II constant repair rate.

With the aid of the Markov method, we write the following equations for the diagram in Figure 8.4 [5, 11–14]:

$$\frac{dP_o(t)}{dt} + (\lambda_1 + \lambda_2 + \lambda_c)P_o(t) - \theta_1 P_1(t) - \theta_2 P_2(t) = 0 \tag{8.40}$$

$$\frac{dP_1(t)}{dt} + (\lambda_2 + \theta_1)P_1(t) - \theta_2 P_3(t) - \lambda_1 P_o(t) = 0 \tag{8.41}$$

$$\frac{dP_2(t)}{dt} + (\lambda_1 + \theta_2)P_2(t) - \theta_1 P_3(t) - \lambda_2 P_o(t) = 0 \tag{8.42}$$

$$\frac{dP_3(t)}{dt} + (\theta_1 + \theta_2)P_3(t) - \lambda_2 P_1(t) - \lambda_1 P_2(t) - \lambda_c P_o(t) = 0. \tag{8.43}$$

At time $t = 0$, $P_o(0) = 1$, $P_1(0) = 0$, $P_2(0) = 0$, and $P_3(0) = 0$.

The following steady-state probability equations are obtained from Equations 8.40 through 8.43 by setting the derivatives with respect to time t equal to zero and utilizing the relationship $\sum_{j=0}^{3} P_j = 1$:

$$P_o = \frac{\theta_1 \theta_2 D}{D_3}, \tag{8.44}$$

where

$$D = D_1 + D_2 \tag{8.45}$$

$$D_1 = \lambda_1 + \theta_1 \tag{8.46}$$

$$D_2 = \lambda_2 + \theta_2 \tag{8.47}$$

$$D_3 = DD_1D_2 + \lambda_c[D_1(D_2 + \theta_1) + \theta_2 D_2] \tag{8.48}$$

$$P_1 = \frac{[D_1\lambda_1 + D_4\lambda_c]\theta_2}{D_3}, \tag{8.49}$$

where

$$D_4 = \lambda_1 + \theta_2 \tag{8.50}$$

$$P_2 = \frac{[D\lambda_2 + D_5\lambda_c]\theta_1}{D_3}, \tag{8.51}$$

where

$$D_5 = \lambda_2 + \theta_1 \tag{8.52}$$

$$P_3 = \frac{D\lambda_1\lambda_2 + D_4 D_5 \lambda_c}{D_3}. \tag{8.53}$$

P_o, P_1, P_2, and P_3 are the steady-state probabilities of the system being in states 0, 1, 2, and 3, respectively.

The steady-state availability of the system is given by

$$AV_s = P_o + P_1 + P_2,   \qquad (8.54)$$

where AV_s is the system's steady-state availability.

8.7   Reasons for improving mining equipment reliability and factors affecting mining system reliability

There are many reasons for improving the reliability of mining equipment. Six of these reasons are as follows [6,15]:

1. *Reason I*: To take advantage of lessons learned from other industrial sectors, such as nuclear power generation, aerospace, and defense
2. *Reason II*: To reduce the cost of poor reliability (note that the true cost of poor reliability in most mining operations, when measured properly, is quite significant)
3. *Reason III*: To provide more accurate short-term forecasts for equipment operating hours
4. *Reason IV*: To maximize profit
5. *Reason V*: To lower the performance of mining equipment–related services in an unplanned manner because of short notice
6. *Reason VI*: To overcome challenges imposed by global competition

There are many factors that affect mining system reliability, either directly or indirectly. Most of these factors are as follows [6,15]:

- *Equipment failure*: This is basically a maintenance-related issue, and it causes an interruption in the production process.
- *Routine maintenance*: In this case, routine servicing, overhauls, and component replacements lead to interruptions in production while the equipment is taken out of service.
- *The blast*: In this case, there is often a need to stop the equipment operation.
- *In effective blasting*: It can result in problems such as poor digability in certain areas and unreliable equipment operation.
- *Accident damage*: It causes an interruption to the ongoing production process if the equipment has to be taken out of service for inspection or repair.
- *Shift changes and crib breaks*: In this case, each and every shift change and crib break generally leads to an interruption to the steady-state nature of the production operation.

- *Spillage and housekeeping*: In this case, the need to stop and clean up spillage around the shovel or the dump area causes an interruption in the ongoing production process.
- *Refueling and lubrication*: In this case, the stoppage of equipment for refueling and lubrication results in the interruption of the ongoing production process.
- *Weather*: In this case, fog or rain can interrupt production.
- *The mine plan*: It generally calls for equipment to be shifted on a periodic basis, as different areas are to be mined. In turn, this results in an interruption to steady-state production.
- *Downstream processes*: In this case, if a downstream process stops in a direct tipping situation, this can lead to an interruption in the mining operation.
- *Minor production stoppages*: In this case, "comfort stops," such as minor adjustments, can interrupt the ongoing production process.
- *Geology*: In this case, variability in digging conditions can result in the need for shovels or trucks to stop. Similarly, different ore grades than the one expected can result in the need to interrupt the ongoing production process.

8.8 Mining equipment reliability measures

There are many mining equipment reliability measures used in the mining industrial sector. Five of these measures are presented next [6,15,16].

1. *Measure I: MTBF*
   MTBF is a commonly used measure, defined by

$$\text{MTBF} = \frac{(H_t - T_d - H_s)}{F_n}, \tag{8.55}$$

where H_t is the total number of hours, T_d is the downtime expressed in hours, H_s is the standby hours, and F_n is the number of failures.

2. *Measure II: Availability*
   Availability is simply the proportion of time the equipment is able to be utilized for its intended purpose. It is defined by

$$AV_e = \frac{(H_t - T_d)(100)}{H_t}, \tag{8.56}$$

where AV_e is equipment availability.

3. *Measure III: Production Efficiency*

Production efficiency may be described simply as the ratio of actual output from a piece of equipment/machine/system (which meets the required quality standards) to its rated output during the period when it is operational. It is defined by

$$E_p = [\{(P_a)/(H_t - T_d - H_s)\}/C_r](100), \qquad (8.57)$$

where E_p is the production efficiency, C_r is the rated capacity expressed in units per hour, and P_a is the actual production.

4. *Measure IV: Utilization*

Utilization is expressed by

$$UT = \frac{(H_t - T_d - H_s)(100)}{(H_t - T_d)}, \qquad (8.58)$$

where UT is the utilization.

5. *Measure V: Overall Equipment Effectiveness*

Overall equipment effectiveness is defined by

$$EE_o = (AV_e)(UT)(E_p), \qquad (8.59)$$

where EE_o is the overall equipment effectiveness.

8.9   *Programmable electronic mining system failures*

Past experiences over the years indicate that various types of hazards can occur with programmable electronic mining system hardware or software failures [6,17]. Hardware failures are physical failures that are generally the result of wear and random events. They can involve any physical element of the system, including programmable electronic devices, data communication paths, sensors, and power supplies. Random hardware failures involve items such as mechanical defects, open circuits, broken wires, short circuits, dielectric failures, and corroded contacts.

In contrast, software failures occur due to systematic (functional) errors that include items such as operator errors, software bugs, timing errors, design errors, management-of-change errors, and requirement errors.

A study of data from the New South Wales Mines in Australia, Queensland Mines in Australia, and the U.S. Mine Safety and Health Administration (MSHA) concerning programmable electronic-based mining systems reported that during the period 1995–2001, a total of 100 mishaps occurred. The breakdown of all these mishaps was as follows [6,17]:

- Systematic failures: 39%
- Random hardware failures: 46%
- Miscellaneous failures: 15%

Both systematic failures and random hardware failures are described next.

8.9.1 Systematic failures

Periodic systematic failures are also referred to as *functional failures*. Some of the sources of these failures are software and hardware design–related errors, errors made during maintenance and repair activities, operator errors, and errors resulting from software modifications.

Analysis of the data for the period 1995–2001 for programmable electronic-based mining systems revealed the following breakdown of the systematic failures [6,17]:

- Maintenance and repair-error-related failures: 40%
- Design-error-related failures: 50%
- Miscellaneous failures: 10%

8.9.2 Random hardware failures

Past experiences over the years indicate that the harsh environmental factors in mines, such as heat, water and dirt intrusion, shock, and vibrations, can significantly influence the occurrence of programmable electronic-based mining system hardware failures. These failures involve items such as solenoids, sensors, power supplies, electrical connectors, and wiring. An example of such failures is the degradation of rubber seals and boots utilized for keeping out moisture and dust.

Analysis of the data for the period 1995–2001 for programmable electronic-based mining systems (as mentioned earlier) revealed the following breakdown of random hardware failures [6,17]:

- Actuator failures: 13%
- Moisture-related failures: 17%
- Electronic-component failures: 26%
- Sensor failures: 33%
- Miscellaneous failures: 11%

Note that switch-related failures are included in the category of sensor failures.

8.10 Methods for measuring winder rope degradation

In underground mines, winders that use steel wire ropes are generally utilized for moving materials. During their usage, the steel wire ropes are subjected to continuous degradation or deterioration. Therefore, the reliability of ropes is extremely important for the performance of shaft-equipped mines, as well as for the safety of the miners involved. Past experiences over the years indicate that the type of damage that occurs in winder ropes during their usage period includes, but is not limited to, corrosion, abrasion formation of loops and kinks, and wire fatigue and resulting breaks [6,18–20]. Due to safety-related concerns, many mining regulatory authorities around the globe mandate that periodic inspections be conducted for determining the condition of winder ropes. Thus, generally two types of inspections (i.e., visual inspection and magnetic nondestructive testing) are conducted, as it is impossible to find rope internal damage and corrosion through visual inspection alone.

Both visual inspection and magnetic nondestructive testing methods are described next.

8.10.1   Visual inspection method

The visual inspection method is considered quite useful to highlight changes in rope diameter and lay length, external wire breaks or loose wires visible corrosion, wear of the crown wires, and any other external damage. In fact, past experiences over the years indicate that visual inspection is the only effective approach for highlighting the degree of severity of rope external abrasive wear, as the magnetic approach tends to underestimate the crown wire wear.

Some of the visual inspection method steps are presented here [6,19,21]:

1. Measure the rope diameter, as well as the lay length at a number of points/sections.
2. Examine for excessive crown wear and broken wires.
3. Examine the entire rope end to end to check for abuse or other damage.
4. Examine the rope termination for the condition of fastening, broken wires, and corrosion.
5. Examine the sheaves for misfit, wear, and so on.
6. Examine the drum condition in the case of drum winders.
7. Check for appropriate lubrication.

8.10.2   Nondestructive testing method

Generally, electromagnetic or permanent magnet–based methods are fairly effective at detecting damage anywhere within a rope in both exterior and interior wires [6,19,22,23]. Magnetic rope testing is conducted by passing the rope through a permanent magnet–based instrument/device. In this case, a length of rope is totally magnetized as it passes through the test device. Magnetic rope testing devices are employed for monitoring two distinct types (i.e., Types A and B) of magnetic field changes due to the existence of anomalies.

Type A is referred to as *loss of metallic area (LMA)*. It may simply be expressed as a relative measure of the amount of material mass missing from a cross section in a wire rope. LMA is measured by comparing a section's magnetic field intensity with that of a reference section on the rope that represents the maximum, unworn metallic cross-section area. Type B changes involve a magnetic dipole generated by a discontinuity in a magnetized section of the rope, such as a corrosion pit, a groove worn into a wire, or a wire break. Often, these are referred to as *leakage flux flaws*.

8.11   Dump-truck tire reliability and the factors affecting their life

In open-cast mines, the shovel-truck system is probably the most flexible one. Its reliability and availability are the most important factors in successfully meeting the open-cast mines' production target. The dump trucks' tires are an important element of the shovel-truck system. Their reliability directly or indirectly affects the open-cast mines' overall production performance.

A study of dump-truck tires revealed that their times to failure followed a normal distribution [24]. It simply means that the tires' failure rate is nonconstant, and their reliability must be computed by using the normal distribution representing the tire failure times.

There are many factors that can affect the life of a dump-truck tire. Some of these are as follows [6,24,25]:

- *Heat generation*: As rubber is a poor conductor, the heat generated through flexing will lead to heat accumulation. Thus, it is to be noted that, although the recommended load-speed inflation pressure ensures an equilibrium between heat generated and heat dissipated, any deviation can result in high temperature.
- *Overinflation*: This reduces the amount of tread in contact with the ground, as well as making tires vulnerable to factors such as impact fractures, cuts, and snags.
- *Underinflation*: This can result in excessive flexing of sidewalks and an increase in the internal temperature of the tire. In turn, this can cause permanent damage to tires, such as casing breakup, ply separation, and radial cracks.
- *Speed and haul length*: It means speed > 30 km/h and a haul length > 5 km can considerably affect the tire's life.

- *Tire bleeding*: Lowering tire pressure after a long run is a normal practice, but it causes premature failures, increases tire temperature, and makes an underinflated tire withstand high load (i.e., vehicle weight).

8.12   Open-pit-system reliability

Nowadays, the proper selection of equipment for modern open-pit mines has become a very challenging issue in terms of reliability, availability, productivity, and othr factors. The overall system has grown to the level where the application of reliability principles has been conceived to be very useful for meeting the ever-growing technological-related requirements. The system is made up of various types of dumping and loading machinery, which in turn are arranged in arrays that result in various types of sequencing systems. The malfunction of a single element in a sequence can cause some of the total amount of system failure.

Nonetheless, each element of the open-pit mine (i.e., working face, dumper, dumping point, shovel, etc.) may be considered as an independent link to the open-pit mine chain system. The chain system's reliability analysis when it forms series and parallel networks are presented next [6,26].

8.12.1   Open-pit-series system

In this case, the open-pit-mine components and units (i.e., shovel, dumper, pumping place, and working place) form a series network. This means that all system components and units must operate normally for the system to succeed. Thus, for independent components and units, the system reliability is expressed by [6,16,26]:

$$R_{oss} = R_{dp}R_{dt}R_{sh}R_{wf},\qquad(8.60)$$

where R_{oss} is the open-pit-series system reliability, R_{dp} is the dumping-place reliability, R_{dt} is the dump-truck/dumper reliability, R_{sh} is the shovel reliability, and R_{wf} is the working-face reliability.

For constant failure rates of the shovel, dumping place, dump-truck/dumper, and working-face, with the aid of Chapter 3 and Reference 16, we obtain

$$R_{dp}(t) = e^{-\lambda_{dp}t}\qquad(8.61)$$

$$R_{dt}(t) = e^{-\lambda_{dt}t}\qquad(8.62)$$

$$R_{sh}(t) = e^{-\lambda_{sh}t}\qquad(8.63)$$

$$R_{wf}(t) = e^{-\lambda_{wf}t},\qquad(8.64)$$

where $R_{dp}(t)$ is the dumping-place reliability at time t, $R_{dt}(t)$ is the dump-truck/dumper reliability at time t, $R_{sh}(t)$ is the shovel reliability at time t, $R_{wf}(t)$ is the working-face reliability at time t, λ_{dp} is the dumping-place constant failure rate, λ_{dt} is the dump-truck/dumper constant failure rate, λ_{sh} is the shovel constant failure rate, and λ_{wf} is working-face constant failure rate.

By inserting Equations 8.61 through 8.64 into Equation 8.60 we obtain

$$R_{oss}(t) = e^{-\lambda_{dp}t}e^{-\lambda_{dt}t}e^{-\lambda_{sh}t}e^{-\lambda_{wf}t}, \tag{8.65}$$

where $R_{oss}(t)$ is the open-pit series system reliability at time t.

By integrating Equation 8.65 over the time interval $[0,\infty]$, we get the following expression for the open-pit-series system MTTF [6,16]:

$$MTTF_{oss} = \int_{0}^{\infty} R_{oss}(t)dt = \frac{1}{\lambda_{dp} + \lambda_{dt} + \lambda_{sh} + \lambda_{wf}} \tag{8.66}$$

where $MTTF_{oss}$ is the open-pit series system MTTF.

EXAMPLE 8.3

Assume that an open-system is composed of a dumping place, dump-truck/dumper, shovel, and working-face, which form a series network and the constant failure rates of these four components are 0.004 failures/h, 0.005 failures/h, 0.006 failures/h, and 0.007 failures/h, respectively.

Calculate the open-pit series system reliability for a 20-h mission and MTTF.

By inserting the given data values into Equation 8.65, we obtain

$$R_{oss}(20) = e^{-(0.004)(20)}e^{-(0.005)(20)}e^{-(0.006)(20)}e^{-(0.007)(20)}.$$
$$= 0.6440$$

Similarly, by putting the specified data values into Equation 8.66, we get

$$MTTF_{oss} = \frac{1}{(0.004) + (0.005) + (0.006) + (0.007)}.$$
$$= 45.45 \text{ h}$$

Thus, the reliability and MTTF of the open-pit-series system are 0.6440 and 45.45 h, respectively.

8.12.2   Open-pit-parallel system

An open-pit-parallel-system network is formed when more than one unit of an open-pit system's units or components function simultaneously, and at least one of these units must operate normally for the system to achieve success. For example, there are two shovels functioning simultaneously, and at least one of the shovels must function normally for system success. In this case, both shovels form a parallel system, and the network's reliability, if both shovels fail independently, with the aid of Chapter 3 and Reference 16, is expressed by

$$R_{psh} = 1 - (1 - R_{sh1})(1 - R_{sh2}), \tag{8.67}$$

where R_{psh} is the parallel network reliability of the two shovels, R_{sh1} is the reliability of shovel number one, and R_{sh2} is the reliability of shovel number two.

For constant failure rates of the shovels, with the aid of Chapter 3 and Reference 16, we write

$$R_{sh1}(t) = e^{-\lambda_{sh1}t} \tag{8.68}$$

and

$$R_{sh2}(t) = e^{-\lambda_{sh2}t}, \tag{8.69}$$

where $R_{sh1}(t)$ is the reliability of shovel number one at time t, $R_{sh2}(t)$ is the reliability of shovel number two at time t, λ_{sh1} is the constant failure rate of shovel number one, and λ_{sh2} is the constant failure rate of shovel number two.

By substituting Equations 8.68 and 8.69 into Equation 8.67 we obtain

$$\begin{aligned} R_{psh}(t) &= 1 - (1 - e^{-\lambda_{sh1}t})(1 - e^{-\lambda_{sh2}t}) \\ &= e^{-\lambda_{sh1}t} + e^{-\lambda_{sh2}t} - e^{-(\lambda_{sh1}+\lambda_{sh2})t}, \end{aligned} \tag{8.70}$$

where $R_{psh}(t)$ is the parallel network reliability of the two shovels at time t.

By integrating Equation 8.70 over the time interval $[0,\infty]$, we get the following equation for the parallel network MTTF for the two shovels:

$$\begin{aligned} \text{MTTF}_{psh} &= \int_0^\infty R_{psh}(t)dt \\ &= \frac{1}{\lambda_{sh1}} + \frac{1}{\lambda_{sh2}} - \frac{1}{(\lambda_{sh1} + \lambda_{sh2})}, \end{aligned} \tag{8.71}$$

where MTTF_{psh} is the parallel network MTTF of the two shovels.

EXAMPLE 8.4

Assume that an open-pit system has two independent and nonidentical shovels that form a parallel network (i.e., at least one shovel operates normally for the system success). The constant failure rates of shovels 1 and 2 are 0.006 failures/h and 0.004 failures/h, respectively. Calculate the reliability of the shovel parallel system/network for a 300-h mission and MTTF.

By inserting the given data values into Equation 8.70, we obtain

$$R_{psh}(300) = e^{-(0.006)(300)} + e^{-(0.004)(300)} - e^{-(0.006+0.004)(300)}$$

$$= 0.4167.$$

Similarly, by substituting the given data values into Equation 8.71, we obtain

$$\text{MTTF}_{psh} = \frac{1}{0.006} + \frac{1}{0.004} - \frac{1}{(0.006 + 0.004)}$$

$$= 316.66 \text{ h}.$$

Thus, the reliability and MTTF of the open-pit parallel system are 0.4167 and 316.66 h, respectively.

PROBLEMS

1. Describe LOLP and difficulties experienced with its usage.
2. Define the following terms:
    i. Power system reliability
    ii. Forced outage rate
    iii. Forced outage
3. Define the following two indexes:
    i. SAIFI
    ii. CAIFI
4. Assume that the constant failure and repair rates of a generator unit are 0.005 failures/h and 0.007 repairs/h, respectively. Calculate the generator unit's steady-state availability.
5. Prove Equations 8.18 through 8.20 by using Equations 8.15 through 8.17.
6. What are the main reasons for improving mining equipment reliability?
7. Discuss at least 10 factors that affect mining equipment reliability.
8. Define the following two measures used in the mining industrial sector to measure equipment reliability:
    i. Production efficiency
    ii. Availability

9. Discuss programmable electronic mining system failures.
10. Describe methods used to measure winder rope degradation.

References

1. Billinton, R., and Allan, R. N., Reliability of Electric Power Systems: An Overview, in *Handbook of Reliability Engineering*, edited by H. Pham, Springer-Verlag, London, 2003, pp. 511–528.
2. Layman, W. J., Fundamental Consideration in Preparing of a Master System Plan, *Electrical World*, Vol. 101, 1933, pp. 778–792.
3. Smith, S.A., Service Reliability Measured by Probabilities of Outage, *Electrical World*, Vol. 103, 1934, pp. 371–374.
4. Dhillon, B. S., *Power System Reliability, Safety, and Management*, Ann Arbor Publishers, Ann Arbor, MI, 1983.
5. Dhillon, B. S., *Applied Reliability and Quality: Fundamentals, Methods, and Procedures*, Springer-Verlag, London, 2007.
6. Dhillon, B. S., *Mining Equipment Reliability, Maintainability, and Safety*, Springer-Verlag, London, 2008.
7. Kueck, J. D., Kirby, B. J., Overholt, P. N., and Markel, L. C., Measurement Practices for Reliability and Power Quality, Report No. ORNL/TM-2004/91, June 2004. Available from the Oak Ridge National Laboratory, Oak Ridge, TN.
8. Endrenyi, J., *Reliability Modeling in Electric Power Systems*, John Wiley and Sons, New York, 1978.
9. Kennedy, B., *Power Quality Primer*, McGraw-Hill, New York, 2000.
10. Billinton, R., and Allan, R. N., *Reliability Evaluation of Power Systems*, Plenum Press, New York, 1996.
11. Dhillon, B. S., and Singh, C., *Engineering Reliability: New Techinques and Applications*, John Wiley and Sons, New York, 1981.
12. Dhillon, B. S., *Reliability Engineering in Systems Design and Operation*, Van Nostrand Reinhold Company, New York, 1983.
13. Gangloff, W. C., Common Mode Failure Analysis, *IEEE Transactions on Power Apparatus and Systems*, Vol. 94, February 1975, pp. 27–30.
14. Billinton, R., Medicherla, T. L. P., and Sachdev, M. S., Common-Cause Outages in Multiple Circuit Transmission Lines, *IEEE Transactions on Reliability*, Vol. 27, 1978, pp. 128–131.
15. Dunn, S., Optimizing Production Scheduling for Maximum Plant Utilization and Minimum Downtime: The Reliability Revolution, *presented at the Dollar Driven Mining Conference*, Perth, Australia, July 19997. Available online at http://www.plantmaintenance.com/ ops.shtml.
16. Dhillon, B. S., *Design Reliability: Fundamentals and Applications*, CRC Press, Boca Raton, FL, 1999.
17. Sammarco, J. J., *Programmable Electronic Mining Systems: Best Practices Reccomendatons (in Nine Parts)*, Report No. IC 9480, (Part 6: 5.1 System Safety Guidance), National Institute for Occupational Safety and Health (NIOSH), US Department of Health and Human Services, Washington, DC, 2005.

Available from the NIOSH: Publications Dissemination, 4676 Columbia Parkway, Cincinnati.

18. Chaplin, C. R., Failure Mechanisms in Wire Ropes, *Engineering Failure Analysis*, Vol. 2, 1995, pp. 45–57.

19. Kuruppu, M., Methods and Reliability of Measuring Winder Rope Degradation, *Mine Planning and Equipment Selection*, April 2003, pp. 261–266.

20. Chaplin, C. R., Hoisting Ropes for Drum Winders: The Mechanics of Degradation, *Mining Technology*, 1994, pp. 213–219.

21. ASTM E157-93, *Standard Practice for Electro-magnetic Examination of Ferromagnetic Steel Wire Ropes*, American Society of Testing and Materials (ASTM), Philadelphia, 1993.

22. Poffenroth, D. N., Procedures and Results of Electromagnetic Testing of Mine Hoist Ropes Using the LMA-TEST Instruments, *Proceedings of the OIPEEC Round Table Conference*, September 1989, pp. 17–21.

23. Weischedal, H. R., The Inspection of Wire Ropes in Service: A Critical Review, *Material Evaluation*, Vol. 43, No. 13, 1985, pp. 1592–1605.

24. Dey, A., Bhattacharya, J., and Banerjee, S., Prediction of Field Reliability for Dumper Tires, *International Journal of Surface Mining Reclamation and Environment*, Vol. 8, 1994, pp. 23–25.

25. Balen, O., Off the Road Tires: Correct Selection and Proper Maintenance, *Journal of Mines, Metals and Fuels*, Vol. 22, 1979, pp. 107–113.

26. Mukhopadhyay, A. K., Open Pit System Reliability, *Journal of Mines, Metals and Fuels*, August 1988, pp. 389–392.

chapter nine

Usability engineering life-cycle stages and important associated areas

9.1 Introduction

Usability engineering is not a one-time effort in which all the user interface–related issues are clearly settled prior to releasing the product or system. In fact, it is a set of activities that normally take place throughout the product/system's life cycle. However, a significant number of these activities are performed during early stages of system/product design, prior to the design of the user interface [1].

Nonetheless, the usability engineering life cycle itself is composed of many stages, and attaining a clear understanding of all of them is essential in order to build user-friendly or effective products and systems. Furthermore, there are many usability-related areas that play a key role in the usability engineering life cycle while producing user-effective products and systems. These areas include fundamental features of design for usability, usability-related actions during product/system design phases, factors affecting usability within organizations, and nonfunctional system development requirements and their impact on usability. This chapter presents several major aspects of usability engineering life-cycle stages and of important associated areas.

9.2 Usability engineering life-cycle stages

Over the years, professionals involved with usability have divided the usability engineering life cycle into many stages [1–6]. For our purposes, it is divided into the following distinct stages [1,7]:

- *Stage I*: Knowing about all potential users
- *Stage II*: Carrying out competitive analysis
- *Stage III*: Establishing appropriate usability goals
- *Stage IV*: Carrying out parallel designs
- *Stage V*: Carrying out participatory design

- *Stage VI*: Coordinating the entire user interface
- *Stage VII*: Applying guidelines
- *Stage VIII*: Prototyping
- *Stage IX*: Carrying out interface evaluation
- *Stage X*: Performing iterative design
- *Stage XI*: Obtaining data from actual field applications

All of these stages are described in the next subsections.

9.2.1   Stage I: Knowing about all potential users

The first stage of the usability engineering life cycle is concerned with finding out about all the potential users of the product under consideration. The two factors that have greatest impact on the usability of product are individual user characteristics and variability in tasks. The individual user characteristic includes items such as the product users' age, educational level, work environment, and work experience. The variability in tasks is basically concerned with the anticipated changes in task-related requirements (i.e., from one extreme to another). For determining the variations in task-related requirements, it is vital to perform task analysis. The performance of task analysis can result in information such as the items presented here [7]:

- The identification of steps involved for carrying out the desired tasks
- A list of tasks that potential users want to perform using the product under consideration
- All of the information the potential user will need to carry out the desired tasks
- The reports to be produced
- The communication-related requirements of users as they carry out the tasks or preparing to carry out the tasks in question

9.2.2   Stage II: Carrying out competitive analysis

Stage II is concerned with analyzing all competing products with regard to their usability, and it is easy to study them from the usability standpoint. For the product under consideration, the information on their usability-related strengths and weaknesses can be very useful. In situations when there are many competing products, a comparative analysis of their user-interface issues can be conducted.

All in all, competitive analysis is very useful for obtaining information on issues that work best, as well as on those that should be avoided as much as possible.

9.2.3   Stage III: Establishing appropriate usability goals

Stage III is concerned with developing appropriate usability-related goals to pursue. Past experiences over the years clearly indicate that it is reasonably easy to develop usability-related goals for new versions of existing products or for products with clearly defined competitors in the market. In such scenarios, the minimum acceptable usability-related goals generally are equivalent to the existing usability levels. On the other hand, when a totally new product is being developed without any competition whatsoever, the establishment of usability-related goals is much more cumbersome.

9.2.4   Stage IV: Carrying out parallel designs

Stage IV is concerned with carrying out the preliminary design of same issues by a number of different designers simultaneously or in parallel. Eventually, the most suitable design is selected, and then is subjected to more detail usability-associated activities. It is to be noted with care that for the best end results, all involved design personnel must work independently.

9.2.5   Stage V: Carrying out participatory design

Although Stage I is concerned with knowing about all the potential product users, at that stage, these users cannot provide effective input on all involved issues. Therefore, during the actual design process, it is very important that all the involved designers have proper access to a pool of representative users.

All in all, the participation of users during the actual design process can help eliminate many usability-associated problems because potential users raise questions that product designers and their associate personnel may not have thought of asking.

9.2.6   Stage VI: Coordinating the entire user interface

Note that consistency is one of the most important characteristics of usability. Therefore, it must apply effectively across the entire user interface, which is what Stage VI is concerned with. Past experiences over the years clearly indicate that to have the entire user interface be consist, it is essential, for each development project, to have a centralized authority that coordinates various activities of the user interface.

Furthermore, note that consistency is not just measured at a single point in time, but rather applies over all subsequent releases of a product. Finally, the more consistent the entire user interface, the less chances for the occurrence of user-related problems.

9.2.7 Stage VII: Applying guidelines

Stage VII is concerned with applying appropriate guidelines that contain well-known principles for user interface design in the development project. There are many different types of guidelines, which are either general or specific.

The main difference between usability-related standards and guidelines is that standards specify how the product interface should appear to potential users, whereas guidelines provide advice about the usability characteristics of the product interface. A number of general user interface guidelines are available in References 8–10.

9.2.8 Stage VIII: Prototyping

In Stage VII, prototypes of the final products or systems are developed for evaluating early usability. A *prototype* may simply be described as a form of design specification that often is used to communicate the final version of the design to the developer. Finally, it is added that the basic idea for prototyping is to save money and time for developing something that can be tested with actual users.

9.2.9 Stage IX: Carrying out interface evaluation

Stage IX is concerned with evaluating the user interface in regard to usability. Over the years, many methods have been developed for testing user interface and their application depends on various factors including the preference of the testing individual. Although every effort should be made for using the most effective test method, the benefits of using some reasonable method for evaluating a user interface rather than releasing it without any evaluation, are much higher than the incremental advantages of employing the exact method. In Chapter 10, the subject of usability testing is discussed in quite detail.

9.2.10 Stage X: Performing iterative design

A new version of the user interface design can be generated during Stage X, done by carefully examining usability-associated difficulties and opportunities, clearly highlighted by empirical testing. However, it should be noted that during the iterative design process, it may not be possible to test all the successive design versions with real potential users. Nonetheless, iterations are a fairly good approach for evaluating design-related ideas simply by trying them out in a concrete design.

Subsequently, the design can be passed on for examination by expert users, usability experts, or consultants, can be subjected to go through heuristic analysis, or both. All in all, note that past experiences over the years, clearly indicate that for achieving maximum benefits, it is essential to plan for multiple iterations.

9.2.11   Stage XI: Obtaining data from actual field applications

After the product is put into actual field use, a variety of usability-related data are collected during Stage XI. The main objectives of collecting such data are to carry out comparisons of the actual user-interface performance with the predicted one, and also to utilize these data in designing the next version of user interface and future products.

9.3   Fundamental features of design for usability and usability-related actions during system design phases

Note that this section basically considers the same elements as Section 9.2, but from a very different perspective. In this case, the five fundamental features of design for usability are shown in Figure 9.1 [4].

These features are described here:

1. *User-centered design*: This is focused, right from the start, on users and tasks. It simply means that all involved designers must learn about all potential product users and the tasks that they perform. This takes place before the start of the design process, and design for usability is initiated by creating a usability-specification document.

2. *Experimental design*: This means that early in the development process, the potential users of the product under consideration should carry out pilot trials using simulations and prototypes. Whenever possible, the alternative versions of pivotal features and interfaces are also simulated or prototyped for conducting comparative usability-associated

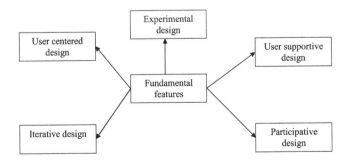

Figure 9.1   Five fundamental features of design for usability.

studies. At the end, assessments are carried out with regard to ease of usability and learning, as well as related difficulties.

3. *Participative design*: This calls for the effective participation of potential users as members of the design team. The effective participation of these users is very important during the design's early formulation phase, as well as when creating the usability specifications. In order for these users to provide useful inputs, they will need to be shown a range of alternatives and possibilities by means of mock-ups and simulations.

4. *Iterative design*: This calls for conducting tests, measures, and redesigns as a regular cycle until the usability specification is fully satisfied.

5. *User-supportive design*: This feature is generally left until a very late stage in the design process and concerns with user-support items such as manuals, training, and help systems.

For usability-related actions, the system design can be divided into five phases, as shown in Figure 9.2 [4].

The usability-related actions taken during each of these five phases are presented next.

1. *Feasibility phase*: Some of the usability-related actions that belong to this phase are as follows:
   a. Conducting preliminary function and operation analyses
   b. Establishing appropriate usability-related goals
   c. Defining the range of potential users of the product under consideration

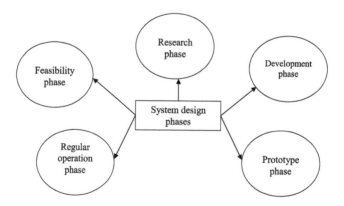

Figure 9.2 System design phases.

    d. Determining if all the requirements are fully satisfied by proposals

    e. Creating and formulizing the usability-related specification

    f. Defining the range of environments and tasks

2. *Research phase*: Three of the usability-related actions taken during this phase are presented here:

    a. Using pilot studies in the field for determining the operational needs of potential users

    b. Conducting studies of human capabilities in regard to system operational-related concepts

    c. Utilizing pilot studies in the field for determining all possible effects upon social and organizational structure

3. *Development phase*: Some of the usability-related actions taken during this phase are as follows:

    a. Examining design ideas against available human-related data

    b. Conducting analysis of tasks and functions that involve humans

    c. Carrying out iterative design

    d. Designing human factor–related aspects of the workplaces and equipment involved

    e. Developing appropriate user-training schemes

    f. Testing subsystem sections with samples of most likely users

4. *Prototype phase*: Three of the usability-related actions taken during this phase are presented here:

    a. Carrying out field trials in realistic environments and with actual users.

    b. Performing iterative designs as necessary

    c. Carrying out a thorough laboratory evaluation by using samples of the most likely users

5. *Regular operation phase*: Three of the usability-related actions taken during this phase are as follows:

    a. Collecting usability evaluation–related data

    b. Giving appropriate data to all concerned bodies

    c. Providing user support (i.e., "hot-line" for help, training, etc.)

9.4 Usability-related myths and factors affecting usability within organizations

There are many usability-related myths, including the following [7,11]:

- *Myth 1*: Usability is user interface design.
- *Myth 2*: Usability lengthens development time and increases development-related costs.
- *Myth 3*: Developers who are familiar with guidelines design good user interfaces.

- *Myth 4*: The user interface is just a matter of adding good graphics for making the application appealing.
- *Myth 5*: Usability can be handled in documentation/training/ help services.
- *Myth 6*: Usability testing is not required if the development team members have been involved with the users for a long period of time.
- *Myth 7*: Usability is really just common sense.

In the case of *Myth 1*, in reality, usability is much more than this; it involves a careful identification of potential users of the product, their goals and capabilities, the tasks they must perform for achieving their set goals, and the context in which they will be using the product under consideration. In the case of *Myth 2*, in reality, the application of usability principles can lower cost and time by eliminating altogether or significantly reducing the need for updates and rework. In fact, various industrial studies conducted over the years show that every dollar spent on user-related efforts during the design phase of the product saves approximately \$10 on problem eradification during the product development phase, or \$100 or more in rework-related actions after the release of the product for application. Furthermore, it is also believed that approximately 80% of all maintenance-associated costs are due to overlooked user-related requirements, and the remaining 20% are due to bugs.

In the case of *Myth 3*, in reality, useful guidelines can simplify the design by reducing the look-and-feel-associated decisions, but they address only a small percentage of questions raised during the interface development process. All in all, it is to be noted that the adherence to guidelines during the user interface development process is certainly very useful; however, their limitations also must be considered carefully. In the case of *Myth 4*, in reality, although it is nice to have good graphics for making the application appealing, the performance of a specified task, efficiently and with ease, is also essential.

In the case of *Myth 5*, in reality, it is not wise to depend on documentation, help, and training for overcoming usability-related difficulties. For example, some surveys of potential users with regard to the use of reference materials and documentation reported the following findings [7,11]:

- Around 11% read printed reference documentation, and another 46% skim it.
- Around 7% read online materials, and another 54% skim them.
- Around 9% read the printed user manual/guide, and another approximately 80% skim it.

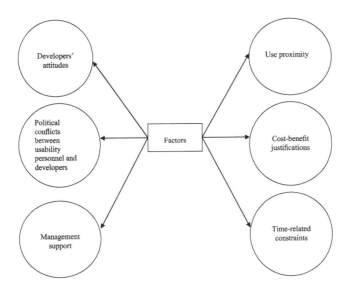

Figure 9.3 Important factors that can affect usability within organizations.

Nonetheless, it does not mean that documentation, training, and help should be ignored altogether; rather, it simply means that complete reliance on them for overcoming usability-related difficulties is not a wise course of action.

In the case of *Myth 6*, in reality, usability testing is always a very good idea, regardless of the length of the association, because changes in user requirements occur quite often. Finally, in the case of *Myth 7*, in reality, the practice of usability definitely requires a proper level of knowledge, experience, and skill.

There are many factors that can considerably affect usability within organizations. The most important of these are shown in Figure 9.3 [12].

9.5 *Nonfunctional system development project requirements and their impact on usability and usability performance measures*

In a system development project, there are many nonfunctional requirements that may affect usability. Six of these requirements, along with their corresponding potential impacts on usability in parentheses, are as follows [7,13]:

1. *Requirement I*: Project size (large projects are more difficult to coordinate and manage and may result in attention on only the most critical usability-related issues).

2. *Requirement II*: Team membership (failure to include usability specialists as members of the team will compromise attention on usability-related matters).
3. *Requirement III*: Portability (the requirement to build multiple compatible versions of a system may result in the development of a "lowest common denominator," and thus lower usability).
4. *Requirement IV*: Economics (the cost of technology of an effective user interface may be too high to include it in the design).
5. *Requirement V*: Legacy systems (earlier commitment to already-existing software/hardware platforms may well over constrain space for design, thus eradicating options that would improve usability significantly).
6. *Requirement VI*: Reliability (a requirement for highly reliable systems may need a complex structure that is reflected later in user interface effectiveness).

There are many usability performance measures, and 12 of the most important of these are as follows:

1. *Measure I*: The number of times the interface misleads the user
2. *Measure II*: The number of times that users need to work around a problem
3. *Measure III*: The number of times that the user loses control of the system
4. *Measure IV*: The number of users who prefer the system
5. *Measure V*: The number of times that the user expresses frustration or satisfaction
6. *Measure VI*: The frequency of help or documentation use
7. *Measure VII*: The number of good and bad features recalled by users
8. *Measure VIII*: The time taken to accomplish a task
9. *Measure IX*: The ratio of successes to failures
10. *Measure X*: The total time spent using help documentation
11. *Measure XI*: The total time spent in errors
12. *Measure XII*: The percentage or number of competitors that do better than the existing product

9.6   Usability advantages

In general, there are many advantages of good usability of products. Some of these are presented here [7]:

- Reduction in maintenance costs
- Reduction in documentation-related costs

- Improvement in efficiency and productivity
- Reduction in support-related costs
- Reduction in training costs
- Improvement in competitive edge
- Reduction in development-related costs
- Better recognition in the media
- Litigation deterrence

The important advantages of usability from the perspective of product users and producers are presented next.

- Users
  - *Less need for support/training*: It simply means that user-friendly products improve learning and can reduce the time that potential users require for seeking support, training, or filtering through relevant documents.
  - *Greater productivity*: It means that user-friendly products allow potential users to efficiently achieve their set goals, rather than wasting time struggling with the product user interface.
  - *Greater satisfaction*: It simply means that potential users will be more satisfied and positive toward a product, rather than being fearful of utilizing it.

- Producers
  - *Reduction in support requirement*: It simply means that most of the time, the user support requirements are due to usability-associated issues, and improvements in usability can lower demands on support.
  - *More competitive product*: It means that easy-to-use items or products are much more competitive and receptive in the market.
  - *Lower cost*: It means that the proper consideration of user needs during the product development phase can lower the need for costly late redesign work, thus reducing overall cost.
  - *Better quality*: It simply means that improving usability will lead to improvement in the perception of users with regard to overall product quality.

PROBLEMS

1. What are the usability engineering life-cycle stages? Discuss at least five of these stages in detail.
2. Discuss fundamental features of design for usability.

3. What are the usability-related actions taken during the following system design phases:
   i. Feasibility phase
   ii. Development phase
4. Discuss at least four myths that are often associated with usability.
5. What are the important factors that can affect usability within organizations?
6. What are the usability-related actions taken during the following system design phases:
   i. Research phase
   ii. Prototype phase
7. List at least eight important measures that can be used to measure usability performance.
8. Discuss the important advantages of usability from the product users' perspectives?
9. Discuss at least four nonfunctional requirements in a system development project that may affect usability.
10. Discuss the important advantages of usability from the product producers' perspectives.

References

1. Neilson, J., *Usability Engineering*, Academic Press, Inc., Boston, 1993.
2. Tyldesley, D. A., Employing Usability Engineering in the Development of Office Products, *The Computer Journal*, Vol. 31, No. 5, 1985, pp. 431–436.
3. Wiklund, M. E., How to Implement Usability Engineering, *Medical Devices and Diagnostics Industry*, September 1993, pp. 68–73.
4. Shackel, B., Usability: Context, Framework, Definition, Design, and Evaluation, in *Human Factors for Informatics Usability*, edited by B. Shackel, Cambridge University Press, New York, 1991, pp. 21–27.
5. Gould, J. D., and Lewis, C., Designing for Usability: Key Principles and What Designers Think, *Communications of the ACM*, Vol. 28, No. 3, 1985, pp. 300–311.
6. Mayhew, D. J., *The Usability Engineering Life Cycle*, Morgan Kaufmann Publishers, San Francisco, 1999.
7. Dhillon, B. S., *Engineering Usability: Fundamentals, Applications, Human Factors, and Human Error*, American Scientific Publishers, Stevenson Ranch, CA, 2004.
8. Mayhew, D. J., *Principles and Guidelines in Software User Interface Design*, Prentice Hall, Inc., Englewood Cliffs, NJ, 1992.
9. Smith, S. L., and Mosier, J. N., *Design Guidelines for Designing User Interface Software*, Report No. MTR-10090, The MITRE Corporation, Bedford, MA, 1986.
10. Brown, C. M. L., *Human-Computer Interface Design Guidelines*, Ablex Publishing Corporation, Norwood, NJ, 1988.

11. Chrusch, M., Seven Great Myths of Usability, *Interactions*, Vol. 7, No. 5, 2000, pp. 13–16.
12. Kaderbhi, T., Overcoming Inertia Within a Large Organization: How to Overcome Resistance to Usability, in *The Politics of Usability: A Practical Guide to Designing Usable Systems in Industry*, edited by L. Trenner and J. Bawa, Springer-Verlag Publishers, New York, 1998, pp. 35–47.
13. Rosson, M. B., and Carroll, J. M., *Usability Engineering*, Morgan Kaufmann Publishers, San Francisco, 2002.

chapter ten

Usability testing and costing

10.1   Introduction

Usability testing is a very important research tool, with its roots going back to classical methodology, and it incorporates a variety of approaches for having potential users try out a product. More clearly, in a typical usability test, the involved users conduct a variety of tasks with a prototype under consideration, while observers monitor their ongoing actions. Testing may involve collecting information on factors such as errors made by users, the degree of user satisfaction with the experience, user frustrations and confusions, and the paths that users take when performing tasks. Thus, *usability testing* may simply be defined as a process that uses participants representing the targeted population to determine the degree to which a product meets specific criteria [1,2].

With regard to usability costing, the cost of each usability engineering–related activities depends on various factors, including the number of users to be studied, scope of the product under consideration, skill and experience of the usability specialists, functionality range, and number of scenarios to be studied. Nevertheless, note that although the cost of usability engineering–related activities may become a significant element of product development cost, past experience indicates that each dollar spent by a manufacturer in developing the usability of product returns $10 to $100 in benefits, in addition to winning the satisfaction of customers and continued business.

This chapter presents various important aspects of usability testing and costing.

10.2   Usability testing goals, limitations, and advantages

The overall goal of usability testing is to highlight and rectify usability-associated shortcomings in products and their associated support materials prior to release. Nonetheless, five important goals of product usability testing are as follows [1,2]:

1. Minimize risk
2. Minimize cost
3. Increase sale

4. Acquire competitive edge
5. Create historical records

The goal "Minimize risk" is concerned with lowering the risk of product malfunction by eradicating serious usability-related problems. The goal "Minimize cost" is concerned with lowering the cost of service and hotline calls. The goal "Increase sale" is concerned with growing sales by making the users of the product happy with it, who will then purchase future releases rather than purchasing the competitors' product.

The goal "Acquire competitive edge" is concerned with gaining competitive traction by convincing a product's potential customers that the product is more user friendly (i.e., easier to use). Finally, the goal "Create historical records" is concerned with developing records of usability benchmarks for use in future products.

There are many limitations and advantages of usability testing. Some of these limitations and advantages are presented in the next subsections [1–4]:

- Limitations
  - User test participants are rarely completely representative of the targeted potential users.
  - It is impossible to simulate exact real-life use environments of the product.
  - There is no 100% guarantee for the usability success of the product.

- Advantages
  - Reduction in the requirement for customer support services
  - Reduction in the learning curve and training-related efforts
  - Accumulation of useful usability-related data for future product releases
  - Wider acceptability of the product by its potential customers
  - Overall improvement in quality of the product

10.3   Usability testing elements, test plans, and test budgets

There are many usability testing elements. The six basic ones are as follows [1,2]:

1. *Element I*: Establishment of test objectives
2. *Element II*: Utilizing a sample of representative users who may or may not be randomly chosen
3. *Element III*: The representation of the real-life use environment

4. *Element IV*: Observing representative consumers who either actually use or are reviewing a representation of the product under consideration
5. *Element V*: Recommending necessary improvements to product design
6. *Element VI*: Data collection, including qualitative and quantitative performance and preference measures

A usability test plan must be appropriately documented well before the start of the testing process. It should address issues such as those listed here [2,5]:

- The place and date of the test
- All test-related goals
- The identification of all test experimenters
- The number of test users needed
- The degree to which experimenters of the test should help test users
- Each test session's length
- The identification of test users and the clear-cut approach to get hold of them
- The user aids to be made available to all test users
- The state of the product at the start of the test
- The identification of data to be collected and the approach for analyzing these data
- The level of computer support needed for the test
- Criteria for determining when users have performed each of the prescribed tasks correctly
- The identification of tasks to be carried out by test users

Generally, a usability test plan includes a budget for the test. Some of the major cost components of a user test budget are shown in Figure 10.1 [2,5,6].

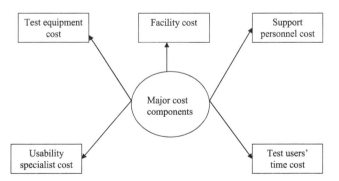

Figure 10.1   Major cost components of a user test budget.

"Test equipment cost" is the estimated cost of equipment to be used during the testing and analysis process. "Facility cost" includes costs other than that of the test equipment cost. The estimated cost of the usability laboratory or other rooms to be used in conducting the test. "Usability specialist cost" is the estimated cost of specialists to be used for planning, running, and analyzing the usability test. "Test users' time cost" is the estimated cost of time to be spent by representative users in testing the product. Finally, "Support personnel cost" is the estimated cost of administrative assistants to be used for scheduling test users, entering data, and so on.

10.4   Types of usability-related tests

Usability-related tests may be divided into the following four distinct types [1,2]:

1. *Type I: Exploratory test*—This test is conducted early in the product development cycle; that is, the product's preliminary defining and designing stages. The main objective of this test is to determine the effectiveness of preliminary design-related concepts with regard to usability. Nonetheless, the test helps to answer questions in areas such as user functions requiring written documentation, user thinking about the product, user interface intuitiveness with regard to operations and navigation, the value of the product's basic functionality to potential users, and prerequisite information needed by potential users.

2. *Type II: Assessment test*—This test is conducted either early or during the middle of the product development cycle, generally after the establishment of the basic or high-level design or organization of the product under consideration. The main goal of the assessment test is to complement the findings of the exploratory test by determining the usability effectiveness of lower-level operations and aspects of the product.

   All in all, unlike the previous test (i.e., the exploratory test), the users of this test always conduct tasks, rather simply walking through and commenting upon items in question. Then, data on quantitative measures are collected, and the interactions between all representative user-participants and monitors are decreased substantially due to greater emphasis on actual user behaviors.

3. *Type III: Validation test*—This test is also known as the *verification test* and is frequently conducted late in the product development cycle. It takes place quite close to the release of the product. The main objective of carrying out this test is to evaluate how well the product under consideration measures up to some predetermined

usability benchmark or standard. Therefore, the intent here is to establish that the product fully meets such a benchmark or standard prior to its release. However, if it fails to meet, then all possible reasons for this scenario need to be established appropriately.

It may be stated that with the exception of three main factors, the validation test is basically conducted in a very similar manner to the assessment test. These factors are focus on quantitative data collection, very little or no interaction between test monitors and user-participants, and existence of benchmarks or standards for the test tasks prior to the start of the test.

4. *Type IV: Comparison test*—This test is not associated with any specific point in the development cycle of a product, but it can be utilized in conjunction with any of the first three test types for comparing two or more alternative designs with regard to usability. Normally, for each alternative design, performance and preference data are collected and a comparison of final results is carried out.

10.5   Usability test performance stages

The performance of a usability-related test may be divided into a number of stages. For our purposes, we have divided it into the following stages [1,2]:

- *Stage I:* Test plan development
- *Stage II:* User-participant selection
- *Stage III:* Test material preparation
- *Stage IV:* Actual test performance
- *Stage V:* Participant debriefing
- *Stage VI:* Data analysis and recommendations

Each of these stages is described in detail in the next subsections [1,2,7–11].

10.5.1   Stage I: Test plan development

Note that although the subject of test plans was discussed earlier in this chapter, it is discussed again here from a different angle, for the sake of completeness. Nonetheless, it may be said that because a test plan is the foundation for the entire test, it must be developed with great care and in a comprehensive manner.

Some of the important reasons for developing a comprehensive test plan are as follows [1,2]:

- To serve as the blueprint for the test
- To provide a focal point for the test
- To describe or imply required resources, both internal and external
- To serve as the main communication vehicle among the main developer, the test monitor, and the rest of the development team members

Note that the format of a test plan may vary from one usability test to another, as well as from one organization to another. However, a typical test plan generally contains sections on topics such as user profile, purpose (general), test monitor role, problem statement/test objective, approach (test design), test environment/equipment, data collection (evaluation measures), and task list [1,2].

10.5.2 Stage II: User-participant selection

The selection and acquisition of representative user-participants for the product under consideration is a very important element of the usability-testing process. The incorrect selection of these people will create final usability-related results that have very little value. The starting point in selecting user-participants is to establish the appropriate user profile of the targeted population by carefully considering factors such as education level; skill and experience (general); demographic-related information; built-in bias toward the product under consideration (negative or positive); the degree of variation in experience, skill, and education levels; and experience and skill regarding the product under consideration.

From time to time, considerable difficulty is experienced during attempts to acquire information concerning the product's users for direct or indirect application during user-participant selection. There are various sources for obtaining such information, including the following typical ones:

- *Marketing manager*: This person is a good source for obtaining user-related information and often has some in-depth analysis with regard to users that has not yet been released to the members of the development team.
- *Product functional specification*: This is the blueprint for the product or item under consideration, which describes all the intended functions of the product or item, in addition to the tasks the users will conduct.
- *Product manager*: This person is also a very good source for obtaining user-related information because he or she generally has access to surveys and reports that describe user profiles in depth.
- *Competitive product analysis group*: In some organizations, this group conducts extensive benchmarking of their own and competitors' products. Thus, it produces good usability-related data not just on company products, but those of competitors as well.

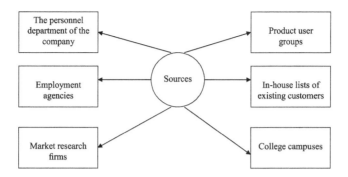

Figure 10.2   Sources for acquiring user-participants for a test.

- *Structured analysis/marketing studies*: These studies are generally conducted by individuals such as human factor specialists, developers, and technical writers prior to the start of any design. They usually contain information concerning the skill and knowledge set needed for using the product effectively.

The process of user-participant selection requires careful consideration when deciding the number of user-participants to test a product. This decision depends on factors such as the availability of the representative user-participants, the time required to prepare for the test, the required degree of confidence in the results, the duration of the testing session, and the amount of resources available to set up and perform the test.

Some of the sources for acquiring user-participants for a test are shown in Figure 10.2 [1,2,11].

10.5.3   Stage III: Test material preparation

Developing the test-related materials for use in communicating with the user-participants, satisfying legal requirements, and collecting data is one of the more labor-intensive activities. The required material's specific contents may vary considerably from one usability test to another. The commonly required materials for developing a test may be grouped under the following categories [2]:

- *Category I: Nonquestionnaire-based*—In this case, the required material items are orientation script, data collection instruments, task scenarios, debriefing topic guide, prerequisite training materials, and nondisclosure agreement.
- *Category II: Questionnaire-based*—In this case, the required material items are background questionnaire, pretest questionnaire, screening questionnaire, and posttest questionnaire.

The items contained in Categories I and II are discussed in the next subsections.

- Category I
    - *Orientation script*: This is a communication tool for user-partici-pants, and its goal is to put user-participants at ease by describing what will happen during the session, in addition to setting the tone for the session in user-participants' minds. Three important guidelines for developing an orientation script are to keep the script tone professional and friendly, to aim to read the script to each user-participant verbatim, and to limit the length of the script to one or two pages, if possible.
    - *Data collection instruments*: These are used for expediting the col-lection of the data necessary to satisfy the test objective effectively. Here, the main intent is to collect these data simply, reliably, and concisely, and the data collected during a usability test falls under one of two broad categories: performance data and preference data.

        Prior to collecting data, proper attention should be placed on factors such as plans to perform data analysis, data collection mechanism, the test plan problems that the data will address, available resources, authority to which data are to be reported, and recording of the data.
    - *Task scenario*: This is a representation of actual work that the user-participants would carry out using the product under consider-ation, and it describes the results to be achieved by the user-par-ticipants, the state of the system at task initiation, actual data and names rather than generalities, and readouts of displays and printouts that will be seen by the user-participants during work performance.

        Some of the important guidelines for developing task scenarios are to make every effort to match the task scenarios to user-partic-ipants' experiences, to provide scenarios that represent real-life situations as much as possible, to arrange task scenarios in the order they are supposed to be executed, and to make every effort to avoid using jargon and cues.
    - *Debriefing topics guide*: It provides structure for carrying out the debriefing session. Broadly, it lists the general topics for discus-sion, and although it suggests a line of questioning, the exact nature of these questions very much depends on the circum-stances surrounding the test session.
    - *Prerequisite training materials*: These are utilized for providing training to user-participants prior to the actual usability test for the purpose of raising their skill to a predetermined level.

- *Nondisclosure agreement*: It is used to prevent the unauthorized disclosures of proprietary-related information about the product in question that user-participants may encounter during the usability test.
- Category II
  - *Background questionnaire*: This is used for collecting historical data concerning user-participants that can be used to understand their behavior and performance during the test. The questionnaire includes a set of questions for obtaining information on the involved user-participants' attitude, preferences, and experience.

    Some of the useful guidelines for developing a background questionnaire are to avoid open-ended questions, to test the questionnaire on an individual fitting the user profile, and to collect all essential information that may affect user-participant performance.
  - *Pretest questionnaire*: This questionnaire addresses specific test objectives, such as to qualify user-participants for inclusion in a specific test group, to determine the user-participants' prerequisite knowledge prior to using the product under consideration, and to ascertain user-participant attitudes and initial impressions about an item's ease of use just before using it.
  - *Screening questionnaire*: This is the vehicle for qualifying and selecting user-participants to take part in the test, and it can be simple or complex, depending on the variability and background of the potential user-participants.

    Normally, the screening questionnaire is presented over the phone to potential user-participants, and some of the useful guidelines for developing a screening questionnaire are to organize questions in a specific order, to review the user/market profile with care, to test the questionnaire on colleagues or others and make revisions as necessary, and to focus on characteristics unique to the product under consideration.
  - *Posttest questionnaire*: This is used for collecting preference-related data from user-participants in order to get a better understanding of the product's limitations and strengths. Some of the useful guidelines pertaining to the posttest questionnaire are to emphasize simplicity in designing questions and responses, to conduct appropriate pilot tests and refine the questionnaire as necessary, to use problem statements from the test plan as the basis for questionnaire content, and to ask questions on issues that cannot be observed directly, such as feelings, opions, and suggestions for improvement.

10.5.4   Stage IV: Actual test performance

Stage IV is concerned with actually carrying out the test; therefore, to do this effectively, careful monitoring is essential. Some of the useful guidelines for monitoring the test are presented here [2,12].

- Consider each new user-participant as an individual.
- Avoid rescuing all involved user-participants whenever they struggle.
- Use humor as much as possible to keep the test session participants relaxed.
- Ensure that all involved user-participants complete their assigned tasks prior to proceedings to the next one.
- Probe and interact with all involved user-participants as the need arises.
- Monitor the test session as impartially as possible.
- Consider using the "thinking aloud" approach, a simple and quite useful method for capturing the thinking of user-participants while working.
- Assist user-participants only in exceptional circumstances.

Additional information on the guidelines for monitoring the usability test is available in References 2,12.

10.5.5   Stage V: Participant debriefing

Stage V is basically concerned with interrogating and reviewing the actions executed by user-participants during the actual performance of the usability test. The basic idea behind debriefing is to comprehend why each error, difficulty, and omission took place during the test. Thus, the debriefing session with all involved user-participants is the last opportunity for fulfilling goals such as these effectively.

The steps of a methodology for carrying out a debriefing session after the main performance test are presented here:

- *Step I*: Explore ideas in regard to debriefing while the involved user-participants complete any posttest questionnaire document.
- *Step II*: Review the completed posttest questionnaire carefully.
- *Step III*: Invite the free-wheeling of thoughts from all involved user-participants.
- *Step IV*: Start questions for general high-level related issues.
- *Step V*: Consider specific issues with care.
- *Step VI*: Review carefully the areas of the posttest questionnaire document completed by the involved user-participants as those that need further exploration.

- *Step VII*: Focus on the comprehension of problems rather than on problem-solving.
- *Step VIII*: Complete the entire line of questioning prior to inviting input from all involved test observers.
- *Step IX*: Leave the door fully open for further contact with all involved user-participants if the need arises.

10.5.6 Stage VI: Data analysis and recommendations

Stage VI is concerned with analyzing the collected data and making necessary recommendations. Four main steps are involved in analyzing the collected usability test data, each of which is described next.

1. *Step I: Compile and summarize collected data*—Compiling is concerned with placing the collected data into a form that allows the comprehension of patterns to a certain degree. Generally, the collected data are summarized using various types of mathematical concepts. For example, the performance data can be summarized under two categories presented next.
2. *Category I: Task timings*—These are related to the time that user-participants require to accomplish each task. Normally, the following statistical concepts are used for summarizing task-timing data values:
   a. Mean time to complete each task (MTTCET)
      This is expressed by

      $$\text{MTTCET} = \frac{\text{TUCT}}{m}, \tag{10.1}$$

      where TUCT is the total of all user-participants' completion times for a given task, and m is the number of user-participants.
   b. Standard deviation of task completion times
      The standard deviation (σ) of user-participants' task completion times for each task is expressed by

      $$\sigma = \frac{\left[\sum T^2 - \frac{\left(\sum T\right)^2}{k}\right]^{1/2}}{k-1}, \tag{10.2}$$

      where T is the time taken by a user-participant to accomplish a task and k is the total number of user-participants' completion times for each task.
   c. *Median time to complete each task*: This is the middle time when the completion times of all involved user-participants are listed in ascending order for a task.

    d. *Range of task completion times*: This comprises the lowest and highest task completion times of all involved user-participants for each task.

3. *Category II: Task accuracy*—There are various ways to determine task accuracy. Three useful indexes that relate to task accuracy are presented here:

    a. *Index I*: Percentage of user-participants conducting their assigned tasks successfully within the specified time benchmark.

    b. *Index II*: Percentage of user-participants conducting their assigned tasks successfully, irrespective of the specified time benchmark.

    c. *Index III*: Percentage of user-participants conducting their assigned tasks successfully, irrespective of the specified time benchmark, including those who needed assistance.

4. *Step II: Analyze collected aata*—Step II is concerned with analyzing the collected data to draw appropriate conclusions. It requires actions such as the following:

    a. Identification of tasks that failed to meet the set criterion

    b. Identification of user errors and related difficulties

    c. Identification of the sources of all errors

    d. Prioritization of all problems by criticality

    e. Analysis of differences between groups or product versions

5. *Step III: Develop appropriate recommendations*—The recommendations created in Step III are very useful for improving the usability of a product. In developing such recommendations, proper attention must be paid to factors such as presented here:

    a. Clear emphasis on solutions expected to have the highest impact on the product.

    b. Clear inclusion of both short-term and long-term recommendations.

    c. Clear identification of areas for further research.

6. *Step IV: Prepare a final report*—The final report generally includes major sections on items such as an executive summary, method, results, findings and recommendations, and appendices. Note that the preparation of a final report is equal in importance to any other crucial usability-associated task. Therefore, documenting it properly is very important.

10.6   Usability cost-related facts and figures

Some of the direct or indirect usability cost-related facts and figures are presented here:

- A study conducted by American Airlines reported that catching a usability-associated problem early in the design of an element could lower the cost of rectifying it by 60%–90% [13].

- A study revealed that the training time required for new users of a standard personal computer was approximately 21 h, as opposed to around 11 h for users of a more usable computer [5].
- In 1991, a study revealed that design changes to usability-related work at IBM resulted in an average decrease of 9.6 min/task, with projected internal savings of $6.8 million [14].
- It costs around $100 billion annually in lost productivity to American businesses because office workers "futz" with their machines an average of about 5.1 h per week [15].
- A study revealed that an $800,000 investment in lowering human factors-related errors for the Line of Sight Forward Heavy resulted in an $80 million cost savings [16,17].
- An Australian study reported that the average cost of user computing across 18 major Australian companies for supporting a single workstation was around $10,000 (Australian), and least 50% of this cost went to "hidden" support (i.e., productivity lost because users stopped their ongoing tasks to help each other with computer-associated problems) [18].
- It is estimated that around 63% of all software-related projects exceed their cost estimates due to factors such as users' lack of understanding of their own needs, frequent requests for changes from users, inadequate user-analyst communication and understanding, and overlooked tasks [19].
- An Australian insurance company spent approximately $100,000 (Australian) on a usability project for redesigning its application forms to make customer errors less likely, saving $536,023 (Australian) annually [20].
- It is estimated that approximately 80% of software maintenance-related costs are due to having to meet unforeseen/unmet user-related needs [21].

10.7   Usability engineering–related activities and costs

A wide range of activities may be utilized to develop a usability engineering product, including the following [2,22,23]:

- Definition of user requirements
- Usability tests (laboratory or field)
- Definition of user profiles
- Specification of usability objectives
- Studies of user work context
- Benchmark studies
- Prototype development (high or low fidelity)

- Surveys and questionnaires
- Task analysis
- Initial design development
- Style guide development
- Thinking-aloud studies
- Prototype redesign
- Paper-and-pencil simulation studies
- Heuristic evaluations
- Focus groups
- Design walkthroughs

The cost of usability engineering for a given product or system includes costs for one or more of these activities. Generally, no more than six of these are accomplished for any one product/system/project [22]. Furthermore, usability engineering–associated work on a project/product/system is tailored to factors such as its requirements, time frame, and resources. Nonetheless, two key guidelines for calculating the usability engineering activities' cost are as follows [22]:

1. *Guideline I*: Ensure that the cost is clearly prorated on the basis of the total number of usability-related tests to be conducted for a stated period, when a permanent usability laboratory is built.
2. *Guideline II*: Ensure that in terms of personnel costs, costs for all development team support, other support, or contract services, as well as all costs associated with participants, are clearly included.

10.8   Costs of ignoring usability and models when estimating usability engineering costs

Past experiences over the years clearly indicate that many products from the standpoint of usability or human factors were poorly designed, but were put on the market anyway and have subsequently required substantial costs to their manufacturers in terms of factors such as tarnished corporate image, reduced sales, and increased customer dissatisfaction. In Reference 24, two examples of scenarios such as these are discussed in detail.

Nonetheless, some of the possible principal costs of ignoring product/system usability are presented here [2,25]:

- *User error cost*: This cost is associated with the users of professional products or systems making errors that lead to reduction in their productivity. The probability of the user error occurrence increases quite significantly if products or systems are difficult to use; thus, there is greater user error cost.

- *Poor sale cost*: This cost is associated with dissatisfied users not purchasing the product or system in future, even after they are made aware of improvements in the usability of the next version. Moreover, it is estimated that a dissatisfied product/system user approximately influence 10 others to avoid buying the product [2,25].
- *Training cost*: This cost is associated with the training of potential users when the product or system is first introduced. It increases significantly for products or systems that are cumbersome to use, whereas for user-friendly products and systems, this cost could be negligible or very small.
- *Tarnish corporate image cost*: This cost is quite difficult to estimate. It occurs when users or customers avoid buying not only the current or improved-usability version of the product/system, but also other products/systems produced by the same company.
- *Poor productivity cost*: This cost is associated with the additional time spent by users of professional products that are difficult to use. It increases significantly if these products are used on a daily basis by a large number of users.
- *Customer support cost*: This cost is associated with a customer hotline telephone service provided by product manufacturers for people experiencing difficulties using the product. Products that are quite difficult to use generate greater user/customer requests for help, and thus they need more people to respond to the users or customers. In turn, the product manufacturer incurs higher customer-support costs.

Many mathematical models can be used to estimate various types of usability engineering-related costs. Four of these models are presented next.

10.8.1 Model I

Model I can be used to estimate the cost of user errors over the entire life-span of a system or product. This cost is expressed by

$$UEC = PLC + ECC, \tag{10.3}$$

where UEC is the user error cost of a system/product over its entire life-span, PLC is the productivity loss cost due to user errors over the system/product lifespan, and ECC is the error correction cost associated with time spent correcting user errors over the system/product life span.

10.8.2 Model II

Model II is concerned with estimating the labor cost of correcting usability problems associated with a system or product for a specified period.

Thus, the labor cost of correcting system/product usability problems is expressed by

$$LC_u = (SSOH)(HLC)\left[\frac{MTTCUP}{MTTUP}\right],\qquad(10.4)$$

where LC_u is the labor cost of correcting system/product usability problems for a specified period, SSOH is the system/product scheduled operating hours for a stated period, HLC is the hourly labor cost of correcting usability-related problems, MTTUP is the mean time to correct system/product usability problems expressed in hours, and MTTCUP is the mean time to correct system/product usability problems expressed in hours.

EXAMPLE 10.1

Assume that a system or product is scheduled to be operated for 5,000 h annually and its mean time to usability problems and mean time to correct usability problems are 500 h and 10 h, respectively. The labor cost of correcting usability problems is $40/h.
Calculate the yearly labor cost of correcting usability problems.
By inserting the specified data values into Equation 10.4, we get

$$LC_u = (5,000)(40)\left[\frac{10}{500}\right]$$

$$= \$4,000.$$

Thus, the yearly labor cost of correcting usability problems is $4,000.

10.8.3   Model III

Model III is concerned with estimating the usability engineering life cycle cost of a system or product. The system/product usability engineering life cycle cost is expressed by

$$ULCC_s = UAC_s + UOC_s,\qquad(10.5)$$

where $ULCC_s$ is the usability engineering life-cycle cost of a system/product and UAC_s is the usability engineering acquisition cost of the system or product. This cost is a component of the procurement cost of a system or product and is associated with usability engineering or human factor–related activities during the system/product design and development phase. Further, UOC_s is the usability engineering ownership cost of the system or product. This cost is an element of the ownership cost (i.e., cost during the operational phase) of a system or product and is concerned with usability engineering-related problems during the system/product operational phase.

10.8.4 Model IV

Model IV is concerned with approximating the usability engineering cost of a system or product when usability engineering cost data are available for similar systems or products of different capacities. Thus, as per Reference 26, the usability engineering cost of the desired system/product can be estimated with the aid of the following equation:

$$UC_{ds} = NUC_s \left[\frac{CP_{ds}}{CP_s} \right]^\theta,\qquad(10.6)$$

where UC_{ds} is the usability engineering cost of the desired system/product; CP_{ds} is the capacity of the desired system or product; NUC_s is the known usability engineering cost of a similar system/product, of capacity CP_s; and θ is the cost-capacity factor whose value varies for different systems/products. In circumstances when no data are available, it is quite reasonable to assume the value of θ to be 0.6.

EXAMPLE 10.2

Assume that the usability engineering cost of a 400-GB hard drive computer is $700. Estimate the usability engineering cost of a 600-GB hard drive computer if the value of the cost-capacity factor is 0.6.
By inserting the specified data values into Equation 10.6, we get

$$UC_{ds} = (700) \left[\frac{600}{400} \right]^{0.6}$$

$$= \$892.8.$$

Thus, the usability engineering cost of the 600-GB hard drive computer is $892.80.

PROBLEMS

1. What are the most important goals of product usability testing?
2. What are the issues that a usability test plan should address?
3. Discuss the types of usability-related tests.
4. What are the sources for acquiring user-participants for a test?
5. Describe the usability test performance stages.
6. What are the possible principal costs of ignoring product usability? Describe these costs in detail.
7. Discuss at least six direct or indirect usability cost–related facts and figures.
8. Write down the equation for estimating usability engineering life-cycle cost of a system.

9. Assume that a system is scheduled to be operated from 4,000 h annually, and its mean time to usability problems and mean time to correct usability problems are 400 h and 5 h, respectively. The labor cost of correcting usability problems is $60/h. Calculate the annual labor cost of correcting usability problems.

10. Assume that the usability engineering cost of a 200-GB hard drive computer is $200. Estimate the usability engineering cost of a 400-GB hard drive computer, if the value of the cost-capacity factor is 0.7.

References

1. Rubin, J., *Handbook of Usability Testing*, John Wiley and Sons, New York, 1994.
2. Dhillon, B. S., *Engineering Usability: Fundamentals, Applications, Human Factors, and Human Error*, American Scientific Publishers, Stevenson Ranch, CA, 2004.
3. Dumas, J. S., and Redish, J. C., *A Practical Guide to Usability Testing*, Ablex Publishing Corporation, Norwood, NJ, 1993.
4. Dumas, J. S., Stimulating Change Through Usability Testing, *SIGCHI Bulletin*, 1989, pp. 37–42.
5. Nielsen, J., *Usability Engineering*, Academic Press, Boston, 1994.
6. Nielsen, J., Usability Testing, in *Handbook of Human Factors and Ergonomics*, edited by G. Salvendy, John Wiley and Sons, New York, 1997, pp. 1543–1568.
7. Zirinsky, M., Usability Testing of Documentation, *IEEE Transactions on Professional Communication*, 1986, pp. 121–125.
8. Mills, C. B., Usability Testing in the Real World, *SIGCHI, Bulleting*, Vol. 19, No. 1, 1987, pp. 43–46.
9. Mills, C. B., and Dye, K. L., Usability Testing: User Reviews, *Technical Communication*, Vol. 32, No. 4, 1985, pp. 40–44.
10. Winbush, B., and McDowell, G., Testing: How to increase the Usability of Computer Manuals, *Technical Communication*, Vol. 27, 1980, pp. 20–22.
11. Dumas, J. S., Conducting Usability Tests, in *Designing User Interfaces for Software*, edited by J. S., Dumas, Prentice-Hall, Englewood Cliffs, NJ, 1988, pp. 25–30.
12. Brown, D. C., How to Get Started in Usability Testing, STC Proceedings, 1989, pp. 165–168.
13. Laplante, A., Put to the Test, *Computer-World*, Vol. 27, July 1992, pp. 75–77.
14. Karat, C. M., Cost-Benefit Analysis of Usability Engineering Techniques, Proceedings of the Human Factors Society 34th Annual Meeting, 1990, pp. 839–843.
15. SBT Accounting Systems, 1997, Westlake Consulting Company, 5444 Westheimer, Unit 1510, Houston, TX.
16. Rouse, W. B., and Boff, K. R., Assessing Cost/Benefits of Human Factors, in *Handbook of Human Factors and Ergonomics*, edited by G. Salvendy, John Wiley and Sons, New York, 1997, pp. 1617–1633.

17. Booher, H. R., and Rouse, W. B., MANPRINT as the Competitive Edge, in *MANPRINT: An Approach to Systems Integration*, edited by H. R. Booher, Van Nostrand Reinhold Company, New York, 1990, pp. 230–245.

18. Ko, C., and Hurley, M., Managing End-User Computing, *Information Management and Computer Security*, Vol. 13, No. 3, 1995, pp. 3–6.

19. Lederer, A. L., and Prasad, J., Nine Management Guidelines for Better Cost Estimating, *Communications of the ACM*, Vol. 35, No. 2, 1992, pp. 51–59.

20. Fisher, P. I., and Sless, D., Information Design Methods and Productivity in the Insurance Industry, *Information Design Journal*, Vol. 6, No. 2, 1990, pp. 103–129.

21. Pressman, R. S., *Software Engineering: A Practitioner's Approach*, McGraw-Hill, New York, 1992.

22. Karat, C. M., A Business Case Approach to Usability Cost Justification, in *Cost-Justifying Usability*, edited by R. G. Bias and D. J. Mayhew, Morgan Kaufmann Publishers, San Francisco, 1994.

23. Mantei, M. M., and Teorey, T. J., Cost-Benefit Analysis for Incorporating Human Factors in the Software Lifecycle, *Communications of the ACM*, Vol. 31, No. 4, 1988, pp. 428–439.

24. Chapanis, A., The Business Case for Human Factors in Informatics, in *Human Factors for Informatics Usability*, edited by B. Shackel and S. J. Richardson, Cambridge University Press, London, 1991, pp. 39–71.

25. Keinonen, T., Mattelmaki, T., Soosalu, M., and Sade, S., *Usability Design Methods, Technical*, Department of Product and Strategic Design, University of Art and Design, Helsinki, 1997.

26. Dieter, G. E., *Engineering Design*, McGraw Hill Book Company, New York, 1983, pp. 324–366.

chapter eleven

Software and web usability

11.1 Introduction

Nowadays, in comparison to first-generation computers, much more money is spent on developing computer software than hardware. For example, in the 1950s, the software component accounted for approximately 20% of the total computer cost, and in the mid-1980s, the percentage increased to about 90% [1]. Today, the annual software industry is worth at least $300 billion in the United States alone [2], and the user interface is often the single most important factor in the success or failure of a software-related project. Furthermore, it is estimated that around 50%–80% of all user source code developed accounts for user interface [3]. User-friendly software enables its users to carry out their tasks easily and intuitively, and it significantly supports quick learning and high skill retention of the involved personnel. All in all, software usability may simply be defined as a quality in software use [3,4].

The World Wide Web (WWW) was released by the European Laboratory for Particle Physics (CERN), and its growth has mushroomed from a mere 623 sites in 1993 to many millions of sites around the globe. Use of the web has become an instrumental factor in the international economy. For example, in 2001, the global e-commerce market was estimated to be about $1.2 trillion, and today it is many trillions. Needless to say, today there are billions of web users around the globe, and the web usability may simply be described as allowing the user to manipulate features of websites for accomplishing a certain goal [3,5].

This chapter presents various important aspects of software and web usability.

11.2 The need to consider usability during software development and software usability engineering

For producing user effective software products, careful consideration to usability during their development process is essential. Some of the important factors that dictate the need for considering usability during the software development process are shown in Figure 11.1. The "Competition" factor means that failure to properly address usability-related issues can result in the loss of market share if competitors release software products with better usability. The "Mixed users" factor means that software

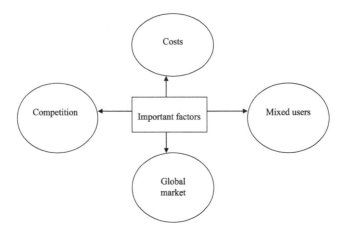

Figure 11.1 Important factors involved in considering usability during the software development process.

product users could be professionals or nonprofessionals with limited or no computer skills at all. The "Global market" factor means that software products generally cover a global market with various language proficiencies, cultures, and other characteristics. Finally, the "Costs" factor means that poor usability software products reduce user productivity and increase the cost of developers in terms of customer support service, hotlines, and so forth.

Past experience over the years clearly indicates that the software usability engineering process may be viewed differently from one organization to another. However, a typical process employed in product development is basically comprised of the following three activities in parallel [6]:

1. *Activity I: Developing an operational usability specification for the software system under consideration*—Usability specification may simply be expressed as a measurable definition of usability that is shared by all involved personnel. It is based on the understanding of the user-related requirements, competitive analysis, and the resources required for producing the software system. Two important points to note with care in developing the usability specification are presented next:
   a. *Point I*: All personnel involved with the development of the usability specification must evaluate it on a continuous basis during the entire development process and make appropriate changes for reflecting up-to-date information on users' requirements.
   b. *Point II*: Failure to understand the requirements of potential users properly prior to developing a specification can result in a specification document that does not reflect those requirements.

2. *Activity II: Adopting an evolutionary delivery approach to software system development*—It means first building a small subset of the software system, and then making it grow throughout the development process, and continuously studying potential users as the system evolves. Furthermore, note that evolutionary delivery exploits, rather than overlooks, the dynamic nature of software-related needs.

Some of the methods employed for improving software system usability during the evolutionary stages are building and testing early prototypes, collecting user feedback during early field tests, implementing a system for collecting usage-related data, and analyzing the impact of design-related solutions [7–9].

Finally, note that an important advantage of the evolutionary delivery approach is that it is very helpful for building the project team members' shared understanding of the software system's user interface design.

3. *Activity III: Visiting customers to understand their requirements*—This is a very important action for gaining insight into ongoing experience of customers with a software system. Data on involved users' experience are collected primarily through contextual interviews (i.e., all relevant interviews are carried out, while users conduct their work activity). During these interviews, the users of the software system are asked about their perception of the system, their work, their system interfaces, and other issues. One significant benefit of these contextual interviews is that they quickly generate large amounts of data.

However, note that different users in different contexts have very different user requirements. The aspects of a user's context that significantly influence the usability of a system for each person include the type of work being conducted, the organization's culture, interaction with other software systems, the physical workplace environment, and the person's social situation.

11.3   Steps for improving usability of software products

The usability of software products can be improved by following the four steps shown in Figure 11.2 [10].

Each of these steps shown in Figure 11.2 is described next.

1. *Step I: Understand product users*—Two important elements involved in understanding product users are the user profile and the usage environment. Knowledge about the user profile helps one to focus design-related efforts on real issues concerning product users and to avoid wasting time and resources on sideline-related issues. However, it requires the collection of information on users' interests, demographics, requirements, and other elements such as education and profession, age range, computer expertise, percentage of male

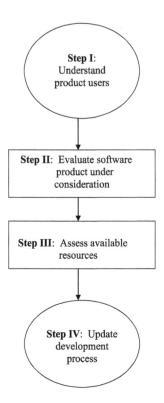

Figure 11.2   Steps for improving software product usability.

versus female, and content knowledge. There are many sources for obtaining such information, including marketing and sales staff; market research data, including surveys and studies; product registrations; and training and customer support staff.

The product usage environment helps one to understand the range of factors that directly or indirectly affect product usage. The environment factors taken into consideration include system configuration, type of network and security, browser types and settings (if applicable), connection speed (if applicable), and location (i.e., mobile, home, office, etc.).

2. *Step II: Evaluate software product under consideration*—Usability is relative. The concept basically means that a software product under consideration may be very easy and usable for one type of users, but quite confusing or intuitive for other users. Thus, an evaluation from the users' perspective involves determining if there is an effective fit between the product and all of its expected potential users.

Nonetheless, a good approach to software product evaluations includes the following actions [3]:

a. *Analyzing the available user data*: This is concerned with performing analysis of the data obtained from sources such as customer service, technical support, marketing personnel, and sales support.

b. *Conducting user field research*: This is concerned with collecting firsthand information on the functioning of the software with its intended users. The best time for conducting user-related field research is at the beginning of the design or redesign project (i.e., to collect information on current usage and on unresolved user needs).

c. *Determining the user-product "fit"*: This is basically concerned with assessing and prioritizing usability-related issues. This should be carried out by keeping in mind that most critical interface elements that affect the success of a software product include screen language clarity, error handling, intuitive navigation, graphic quality, presentation performance, and personalization of content.

d. *Testing usability*: This is concerned with determining exactly where users are having problems in using the software product under consideration. Usability testing can be conducted in many settings, including a designated area within the company establishment and on customer premises.

3. *Step III: Assess available resources*—Step III is concerned with assessing the existing resources, as well as the ability of all personnel involved in executing the software design/redesign project. The diverse talents and capabilities of the involved design team members should be taken into consideration as well. For example, it could be very useful to produce quality screen writing, interaction design, and graphic design of web pages.

Also, it is very important to assess the existence of user advocacy among the involved design team members because it is pivotal for producing an effective, user-centered design. All in all, there should be at least one person among the involved team members clearly advocating the users' point of view.

4. *Step IV: Update the development process*—In order to maintain an effective usability process for future releases or uses, proper evaluation of the overall development cycle is essential. The critical elements of a user-centered design process are establishing appropriate methods of tracking results, establishing cycles of user feedback, and keeping satisfactory design-associated documentation.

Nonetheless, the documents that are considered useful for software design and redesign include functional specification, user interface specification, marketing (business) requirements, flowcharts, and application specification. All in all, the proper evaluation of documents such as these is a very important step toward updating the development process.

11.4   Software usability inspection methods and their selection-related considerations

Software usability inspection methods are used by usability experts to evaluate the software's user interface. There are many usability inspection/evaluation methods [11], including the following [10]:

- *Heuristic evaluation*: It involves usability experts to determine whether each dialogue element properly meets set usability principles. The guidelines checklist helps to ensure that the proper usability-related principles will be considered in software design work. Moreover, the checklist provides inspectors with a basis for comparing the software product under consideration. Generally, checklists are utilized in conjunction with a usability evaluation or inspection method.

    Heuristic evaluation is easy to learn and useful for identifying problems early in the design process, and it also is inexpensive to implement [12]. Its main limitation is that it requires a proper debriefing session to find out how to rectify problems.
- *Cognitive walkthrough*: It uses a detailed procedure for simulating task execution at each step of the dialogue for determining if the simulated user's goals and memory content can be safely assumed to lead to the next anticipated action. The main benefit of this method is that it is an effective tool for predicting problems and capturing cognitive process. On the other hand, it has two main drawbacks: the need to train a skilled evaluator and the fact that it focuses on only one attribute of usability [12].
- *Guidelines checklists*: They help to ensure that the proper usability principles will be considered in software design–related work. Generally, checklists are utilized in conjunction with a usability evaluation or inspection method.
- *Pluralistic walkthrough*: It involves group meetings where usability developers, experts, and students step through a learning scenario and discuss all the dialogue elements. Furthermore, this feature of inspection lists items such as sequences of features utilized for carrying out typical tasks, checks for long sequences, difficult steps, and steps that need extensive experience in order to assess a proposed set of features properly. Some of the main benefits of this method are useful for meeting the criteria of all involved parties, easy to learn and use, and permits iterative testing/evaluation [3,12].
- *Standards inspection*: It involves usability specialists who inspect the interface compliance with stated standards. These standards could be departmental standards (if any), domain-specific software standards, or user interface standards.

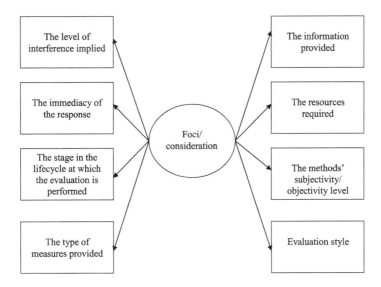

Figure 11.3 Foci/considerations for selecting a proper usability inspection or evaluation method.

In the selection of a proper usability inspection/evaluation method or combination of methods for a specific application, the various foci of the evaluation must be considered with care. Most of these considerations are shown in Figure 11.3 [3,13].

11.5 *Software usability testing methods, and important factors with respect to such methods*

There are many usability testing methods for measuring system or product performance against predefined criteria, as per the usability attributes suggested by the usability standards and metrics [3,14]. Generally, in these methods, users carry out certain tasks with the software system or product. Moreover, data are gathered on measured performance (e.g., the time needed to accomplish a task). Four widely used software usability testing methods are as follows [3,14]:

1. *Method I: Performance measurement*—In this case, usability tests are directed toward determining hard and quantitative data. Mostly, these data take the form of performance metrics (e.g., the required time for performing certain tasks). Note that a usability evaluation method based on measured performance of predetermined usability metrics is strongly promoted by the International Organization for Standardization (ISO) [15].

2. *Method II: Thinking-aloud protocol*—This is a quite commonly used method in software usability testing because it helps to carry out formative evaluation is usability tests [16]. More specifically, during the testing process, all involved participants are invited to express their feelings, thoughts, and opinions while interacting with the software and conducting tasks. These user remarks provide significant insight into the most suitable method of designing the system interaction. However, this method could be difficult to use with some user groups, such as young students, who are distracted by the process.

3. *Method III: Codiscovery*—In this case, a group of users conduct tasks together that simulate a typical work process under observation. Moreover, most of these involved persons have someone else available to help. All in all, in many work scenarios, the method is considered very useful.

4. *Method IV: In-field studies*—This method is concerned with observing users conducting their tasks in a normal work environment. The main benefit of the method is the real-life user performance and group interaction. However, in terms of measuring performance, it has certain limits because appropriate testing equipment cannot be used properly in normal work environments.

All in all, factors such as the following are very important with respect to software usability testing methods [3,17]:

- Selection of representative tasks to be conducted by users
- A well-organized usability laboratory
- Selection of appropriate software users
- A well-designed usability laboratory

11.6   Useful guidelines for conducting software usability testing

Over the years, professionals working in the area of software usability have developed a number of useful guidelines for performing testing. Seven of these guidelines are presented here [3,12]:

1. *Guideline I: Assist the individuals taking part in the usability test only in exceptional circumstances*—More specifically, let these individuals struggle as much as possible and solve their own problems.

2. *Guideline II: Use humor as appropriate for keeping the test environment fairly relaxed*—Humor is very useful for counteracting participants' self-consciousness, and it can help them to relax.

3. *Guideline III: Use the thinking-aloud approach as appropriate*—This method has proved to be very effective for capturing the thinking of the participating personnel as they work with the interactive software.

4. *Guideline IV: Keep the testing session as neutral as possible*—It basically means that there should be no vested interest whatsoever in the results of the test, one way or the other.

5. *Guideline V: Ensure that the person conducting or directing the usability test keeps going even when he or she makes an error*—This individual should not panic even in situations when he or she has inadvertently revealed information or in some other way has biased the usability test in progress. If the test just continues, then the action may not even be observed by the participants.

6. *Guideline VI: Treat each person taking part in the test as a completely new case*—Consider that each test participant is unique, regardless of background or earlier performance in usability testing sessions.

7. *Guideline VII: Ensure that the individual conducting or directing the usability test is highly conscious of his or her body language and voice*—This is very important because it is fairly easy to unintentionally influence someone through body language and voice.

11.7   Web usability–related facts and figures

There are many facts and figures concerned with web usability, whether directly or indirectly, including the following:

- There are over 800 million pages on the web in the United States alone [18].
- A study revealed that 40% of all website users and visitors elected not to return to a site because of design-associated problems [19].
- Each year, web usability is improving somewhere between 2% and 8% [20].
- One study reported that approximately 65% of all online shopping trips end up in failure [21].
- A study of e-commerce sites reported that about 56% of intended tasks were conducted successfully by the site users [22].
- User interface elements account for about 47%–60% of the lines of system or application code [23].
- An average mid-sized company could save approximately $5 million annually in employee productivity by improving the design of its intranet (the system used only by company employees) [24].
- About 10% of users scroll beyond the information that is visible on the screen when a page appears or comes on [25].

11.8   Common web design–related errors

Over the years, professionals working in the area of web design have highlighted many basic errors that are quite common on all levels. Six of these errors are as follows [26]:

1. *Page design error*: This is creating attractive pages for evoking positive feelings when utilized within the company/organization, rather than designing to achieve an optimal user experience in a realistic environment, although the pages are less attractive.
2. *Information architecture error*: This is structuring the website for mirroring the company or organization structure, rather than structuring it for mirroring users' tasks and their views of the information space.
3. *Project management error*: This is managing a web project as a conventional corporate project, rather than managing it as a single customer-interface project. The main drawback of the traditional approach is that it directly or indirectly leads to an internally focused design with an inconsistent user interface.
4. *Content authoring error*: This is writing in the usual linear style, rather than specifically writing for online readers by clearly keeping in mind that these readers quite often scan text and require rather short pages with secondary information relegated to supporting pages.
5. *Linking strategy error*: This is treating a website as indispensable, thus not providing proper links to other sites, as well as not having appropriate entry points for others to link to. Furthermore, many companies and organizations even overlook the application of proper links to their own site in their own advertisements.
6. *Business model error*: This is treating the web as a Marcom brochure rather than a basic shift that will change carrying out business transations in the age of network economy.

11.9   Web page design

The look and feel of a web page is the most immediately visible component of web design, and it is a critical factor in the effectiveness of website usability. Some of the important usability "dos" and "don'ts" with regard to the web page design are presented next [3,27].

- "Dos"
  - Aim to keep the size of most web pages to a level that can be downloaded within 10 s.
  - Design web pages in such a way that the browser can easily resize them for meeting his or her specific needs.
  - Use visual highlighting as appropriate for drawing users' attention to pertinent information.

- Tailor images to elements that are clearly meaningful. More specifically, past experiences over the years clearly indicate that dense graphics alienate many users.
- Aim for fitting main page contents in the potential browser window's width, even in situations when the window is not maximized to fill the entire screen.
- Consider screen real estate as a highly valuable commodity.

- "Don'ts"
  - Avoid assuming that all potential users can clearly see what you see.
  - Avoid using all capital letters.
  - Avoid specifying fonts using absolute sizes.
  - Avoid using animation unless it is absolutely necessary.
  - Avoid getting carried away with artistic or creative fonts.

Some of the most important factors to be considered in web page design are as follows [3,27]:

- Font usage
- Page size
- Image usage
- Help for users
- Textual element usage

Each of these factors is discussed in detail in the next subsections.

11.9.1   Font usage

Fonts are used to create various types of web page elements, such as navigation bars, menus, buttons, headers and footers, links, and tables, in addition to the text that conveys most of a website's contents. Font faces fall into the following two categories:

1. *Sans-serif fonts*: These are simpler in shape because they consist of only primary line strokes. Arial, Futura, and Helvetica are the typical examples of these fonts. Quite often, sans-serif fonts are utilized for short phrases such as button labels, outlines, and titles.
2. *Serif fonts*: These have small appendages at the tops and bottoms of letters. Courier, Times Roman, and Century are the examples of the serif fonts.

Some font usage–associated pointers are as follows:

- Use italics for defining terms or emphasizing an occasional word.
- Avoid getting carried away with the use of font sizes, styles, and faces.

- Different browsers support different font faces.
- Avoid specifying absolute sizes of fonts.

11.9.2   Page size

Page size is an important factor of usability effectiveness. More specifically, it is important to usability in two ways: the downloading and displaying speed of pages, and the flexibility of the pages for fitting the available display area. Both of these are described in the next subsections.

11.9.2.1   Page downloading speed

The length of time it takes to download a page from the server and display it in the browser window is a key factor of sizing web pages properly. *Response time* is defined as the time between when a user requests a page to when it has been displayed totally. Some of the useful guidelines associated with response time are as follows:

- Provide adequate warning to users when a web page will need more than 10 s to download.
- Ensure that the response time is within 10 s in order to hold the users' attention.
- Ensure that the response time is within 0.1 s to make the system "feel interactive."
- Ensure that the response time is within 1 s in order to fit into the user's chain of thoughts.

11.9.2.2   Page flexibility

Page flexibility is an important factor in usability effectiveness. For its successful achievement, the following guidelines can be very useful:

- Ensure that all the key page elements are clearly visible with scrolling when the window is 400 pixels high.
- Design web pages in such a way so that they can easily be resized (i.e., to fit within a fairly wide range of window sizes).
- Pay careful attention to resizeability when designing headers, footers, and other elements.
- Use relative rather than absolute sizes for elements that fall under the browser's resizing capability.
- Normally, no horizontal scrolling is necessary if the window is 800 pixels wide.

11.9.3   Help for users

Generally, users do not read web pages in a serial manner; rather, they shift their attention from one visual element to another. Thus, the biggest challenge faced by web page designers is utilizing the visual elements effectively, in a way that will draw the attention of users to vital content. More specifically, potential users will not read text unless they are properly enticed to do so. Some of the important pointers in this regard are presented here [3,27]:

- Studies clearly indicate that users generally assume that a row of similar elements should be read from left to right or top to bottom.
- Use size to help users understand which elements fall where with regard to the content hierarchy.
- Ensure that the most important text has the largest possible contrast.
- Reinforce the hierarchy of web page content with an emphasis on visual dominance.
- Ensure that the use of visual highlighting methods is consistent across the entire website.
- Use white space for creating distance between groupings, as well as to group related elements together.
- Test every web page's final design by eliminating all the visual elements, one at a time.
- Studies clearly indicate that items above and to the left of the center of a page are noticed first.

11.9.4   Textual element usage

Generally, most of the key elements of a website are conveyed through the use of text, lists, and tables. Therefore, it is extremely important to write in a style that not only transmits desired information, but also reflects how websites are actually employed. Some of the useful guidelines for writing web page text are presented here [3,27]:

- Ensure that subjective or exaggerated language is converted clearly into more neutral terms.
- Ensure that the text layout is converted properly to a format that is more "scannable."
- Ensure that all the text is as concise as possible.
- Combine the first three factors to obtain maximum benefits.

11.9.5   Image usage

The creativity with which images of all types (e.g., photographs, diagrams, and plots) are integrated effectively with textual elements is considered to be a key factor for the popularity of a website. On the other hand, users generally blame images for many of the usability-associated problems that plague web access. Several useful guidelines for using images are presented here [3,26,27]:

- Avoid using a different photograph on each page of a website. Otherwise, the performance of the website will be sacrificed.
- Use a commercial image compression tool to reduce the size of image files as much as possible.
- Use animation only in those circumstances where it truly adds to the meaning of the information being presented.
- Limit graphics to elements that are really necessary.
- Limit as much as possible the number of different colors used on the site.
- Use the most efficient format for each image.
- Aim to include a thumbnail image on the web page that links to the larger one.
- When you employ a graphic, make every effort to reuse it on other pages if the need arises.
- Reduce the image resolution as much as possible.

11.10 Website design

Generally, greater attention is paid to page design than site design. However, from the viewpoint of usability, the site design is much more important and challenging. Some of the key usability "dos" and "don'ts" are presented next [3,27].

- "Dos"
  - Ensure that a single website's pages share a common look-and-feel as much as possible.
  - Ensure that each web page incorporates real content.
  - Ensure that each web page honors the user's browser settings.
  - Ensure that the web page effectively supports browser resizing as much as possible.

- "Don'ts"
  - Avoid having a copyright notice. It is not required for establishing ownership to web material.
  - Avoid using frames.
  - Avoid saying "welcome" on web pages.
  - Avoid having a "banner page."
  - Try not to use pop-up windows without the consent of the involved user.

Three of the key factors to be considered in website design are discussed in the next subsections [3,26,27].

11.10.1 Site testing and maintenance

To maintain usability effectiveness, regular testing and maintenance of websites are essential. Moreover, this is a very useful way to ensure that the users see the site's intended message. To ensure the quality of the web page, test the design and each new page by utilizing at least Microsoft Internet Explorer and Netscape browser, disabling images, utilizing different browser window widths, and utilizing dial-up connections [3,27].

Web pages tend to change over time, and that is another reason why it is very important to conduct maintenance-related activities on a regular basis. The verification of all links still being active must be conducted at least once a month. Moreover, whenever a web page is modified, it is essential to double-check that its links are functioning correctly.

11.10.2 Site organization

Site organization must be carefully considered during the design process because past experience indicates that users do not read web pages in the way they read books. Thus, some useful guidelines concerned with website organization, whether directly or indirectly, are presented here [3,26,27]:

- Ensure that the important information on a page is positioned in such a way that it is clearly visible, even when the browser window is shrunk to around 50% of the screen width.
- Ensure that users are provided with at least some content on each page.
- Organize the site into a number of bite-size pieces, capable of being easily transversed in different ways to take advantage of the web's navigational flexibility.
- Avoid as much as possible displaying blocks of text in a large font.
- Ensure that all pointers to related topics are clearly visible somewhere in the upper half of the page under consideration.
- Avoid requesting potential users to provide their e-mail addresses.

11.10.3 Shared elements of website pages

To achieve effective usability of a site, it is very important to help its users become familiar with the site with minimum effort. This can be achieved only by adopting a consistent page style that clearly repeats common elements throughout the website. This process also can be very helpful to improve user speed. Another factor that is helpful to usability is to concentrate all common elements at the bottom and top of each page, or along the left side.

As per some usability studies, some user expectations of common elements are as follows [3,26,27]:

- Inclusion of a help feature only in those situations that provide substantive useful information.
- Ability to go to the home page by simply clicking on the website's icon.
- Inclusion of a search mechanism only in circumstances where the site includes a very large number of pages.
- Gather of information intended for sponsoring agencies on the "About us" tab.
- Clear displaying of contact information for the website's sponsor or creator under the "Contact us" tab.

11.11   Navigation aids

Users employ navigation aids to find their way around a website. Some examples of the navigation aids are navigation bars, links, and menus. Over the years, many studies concerning navigation aid usability have been performed. Some of the important "dos" and "don'ts" concerning the usability of navigation aids are presented next [3,27,28].

- "Dos"
  - Ensure that all links, navigation bar items, or menu items that lead to the "current location" are deactivated appropriately.
  - Organize all navigation aids by carefully considering the user tasks to be carried out.
  - Give high importance to navigation bars or breadcrumb trails because they provide users with an understanding of where they are on a site.
  - Ensure consistent conformance to the standard practice of having all links underlined.

- "Don'ts"
  - Avoid changing the standard colors used for links.
  - Avoid labeling menu items, links, or buttons with phrases that are meaningless.
  - Avoid confusing users by implementing menus in a creative fashion as much as possible.
  - Avoid assuming that all potential users will be able to familiarize themselves with the website easily.

Three of the important factors to be considered with regard to navigation aids are shown in Figure 11.4.

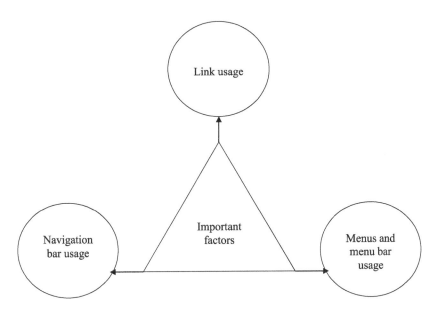

Figure 11.4 Three of the important factors to consider with regard to a navigation aid.

Each of the important factors shown in Figure 11.4 is described in the following subsections [3,27,28].

11.11.1   Navigation bar usage

The main objective of having navigation bars is to lay out a website structure in a hierarchical format. Generally, they are located along one side of the web page. However, sometimes they also may be found in a box at the top of the web page's main content area. There are two key factors that have proved to be very helpful in using the navigation bar properly: (1) identification of the top 10 tasks, during the development of the navigation bar, that potential users are most likely to do on the site; and (2) selection of navigation structure and labels with the utmost care.

The following steps are considered very useful for selecting navigation structure and labels [3,27,28]:

Step I: Develop a list of all the functionalities or operations that the website must support for potential users for accomplishing the highlighted top 10 tasks properly.
Step II: Write down the operations on individual index cards.
Step III: Spread out the cards and group them into logical categories/ classifications.

Step IV: Highlight at least five potential users, and have them repeat Steps I–III.

Step V: Compare the results of all the sorting. If some pattern of classification is not clearly emerging, repeat the entire process again.

Step VI: Once the classifications arrived at by most potential users clearly appear to be similar, use them to develop an outline of the website structure.

Step VII: Present the structure outline to at least five other potential users and ask for their input. Repeat the process as necessary.

11.11.2   Usage of menus and menu bars

Generally, menu bars are utilized by websites for providing basic navigation functionality. These bars may include various links or contain menu titles that drop down as users click on them or pass over them. Some of the important guidelines concerning menu bars and menus are presented here [3,27,28]:

- Aim to anchor all menus to a menu bar across the top of the web page.
- Avoid using cascading (i.e., multilevel) menus as much as possible.
- Ensure that menu titles are short and clearly form a consistent group.
- Format menu titles and menu items by utilizing uppercase and lowercase letters in a standard fashion.
- Ensure that the menu is obviously identifiable as such by making the title look like a typical link.
- Group menu items together as logically as possible.

11.11.3   Link usage

Links are probably the most common mechanism supporting website navigation. Web designers use links in the following three ways [3,27,28]:

1. To direct users to pages that contain additional information regarding the graphics and text mentioned in a link
2. To direct users to an alternative source, if the current page does not contain the needed information
3. To provide efficient access to the website's other pages

Some of the important guidelines concerning the effective use of links are presented here [3,27,28]:

- Underline words that really matter in order to improve link readability.
- Underline all links and use standard colors.

- Avoid underlining nonlinked text as much as possible, as the underscoring will make it look like a link is there.
- Classify the links into categories when multiple links appear in a list.
- Format all links by utilizing uppercase and lowercase letters.
- Make the image itself be the link if there is a need to link a larger copy of an image.
- Choose link text carefully.
- Locate all alternative links at the top of the page.

11.12 Tools for evaluating web usability

There are many tools that can be used to identify potential usability-related problems. These tools and methods are good for checking routine site-design elements with regard to consistency, as well as for encouraging the use of good design practices. Four of these are NetRaker, Lift, WebSAT, and Max. Each of these tools or methods is described in the next subsections [3].

11.12.1 NetRaker

NetRaker is made up of various tools that help to highlight usability-related problems and to conduct market research. Furthermore, it provides a comprehensive set of useful guidelines for composing objective survey questions and a customizable set of usability survey templates. The questions are made randomly available to website users by giving the option to participate. The survey requires users to conduct tasks on the website and then to provide adequate feedback with regard to the ease of carrying out the tasks.

Some of the benefits of NetRaker are presented here [3]:

- It surveys users and gathers usability feedback quickly.
- NetRaker's automation ensures that users are surveyed consistently.
- It is very useful for obtaining feedback in the context of a website's intended purpose, as opposed to fully relying on generic Hypertext Mark-up Language (HTML) checks or statistical analysis.

Finally, NetRaker is one of the best tools for highlighting usability-related issues because it is based on users' direct feedback.

11.12.2 Lift

Lift is another very useful tool that conducts analysis of a web page for potential usability-related problems. There are two types of Lift: Lift Online and Lift Onsite. Lift Online conducts HTML checks that are derived from usability principles in a similar way to Web SAT. More clearly, it examines only one page at a time and then provides a report on the page's usability-related issues. Furthermore, Lift Online goes one step further than Web SAT, in that it also provides specific code change recommendations.

Lift Onsite can be run quite easily from a personal computer (PC), and it provides the compelling feature of directly fixing HTML-related problems as they are being reviewed in the usability evaluation report.

Finally, Lift provides usability-based HTML validation to ensure good coding-related practices.

11.12.3   Web SAT

Web SAT is a web static analyzer tool that belongs to the Web Metrics Suite developed by the National Institute of Standards and Technology (NIST). It is employed for checking web page HTML against typical usability guidelines for potential problems. Web SAT allows up to five individual Uniform Resource Locators (URLs) to be checked against its usability-related guidelines. At the end of the evaluation process, the tool provides a report of problems found on each web page that was entered.

The problems are classified into the following six categories [3]:

1. *Category I: Performance*—Its problems are associated with the size and coding of graphics in relation to page download speeds.
2. *Category II: Accessibility*—Its problems are associated with the page making proper use of tags for visually impaired users.
3. *Category III: Readability*—Its problems pertain to content readability.
4. *Category IV: Navigation*—Its problems are associated with the coding of links.
5. *Category V: Form use*—Its problems are associated with the form's Submit and Reset buttons.
6. *Category VI: Maintainability*—Its problems are concerned with tags and coding information that would make the page easier to port to another server.

All in all, the main limitation of Web SAT is that it can check only individual web pages.

11.12.4   Max

Max is another useful usability tool, and it employs a statistical mode for simulating a user's experience with calculating ratings in the following areas [3]:

- *Content*: In this case, Max summarizes the percentage of various media elements (i.e., text, graphics, and multimedia), as well as client-side technologies utilized such as Portable Document Format (PDF) and Flash, which comprise the website under consideration.

- *Load time*: In this case, Max estimates the average time taken for loading website pages.
- *Accessibility*: In this case, Max estimates the average time that a user takes for finding something on the site.

All in all, the main strength and weakness of Max are that it provides a performance benchmark, but it does not provide many suggestions for making design changes, respectively.

11.13 Questions for evaluating the effectiveness of website message communication

There are many general questions to use to evaluate the effectiveness of website message communication that are considered to be useful in improving web usability directly or indirectly. In this section, these questions are grouped under six distinct areas: Navigation, Content, Design, Text, Mechanics, and Concept. Questions belonging to each of these areas are presented next [3,29–31].

1. Navigation
   a. How easy is it to use the hyperlinks, menu, and buttons for browsing the site effectively?
   b. How does the site conform to current web standards, and does the site have a local search engine?
   c. Will it always be clear to visitors regarding their browsing location?
   d. Is it possible to predict what the contents of the options on the menu are without clicking them?
   e. Is it possible to find anything when using simple and straightforward keywords on the search engine?
   f. Is it possible that potential visitors will ever be bewildered by the appearance of something unexpected?
2. Content
   a. How accurate and unique are the website's contents?
   b. Does the site offer any communication options clearly?
   c. What portion of the web pages was allocated for content, in comparison to other factors?
   d. What type of information is the website expected to convey?
   e. Is the content fairly easily accessible and the purchasing procedure (if applicable) user-friendly?
   f. Is the information on the website easily accessible, accurate, complete, attractive, and clear?
3. Design
   a. Is the site style unique, and the design impressionable?

    b. Is there a proper balance between content and design?
    c. Is there sufficient contrast between text background and color?
    d. Are visitors to the site directed properly to the most important page elements?
    e. Can the design in any way deter visitors?

4. Text
    a. Is the text grammatically checked and sufficiently readable?
    b. Does the first page effectively convey a message to its visitors, including what they can expect to find on the website?
    c. Are the button and hyperlink titles properly clear?
    d. Is the text on the first and featured pages brief enough to use properly?
    e. Are the titles and subheadings informative enough?
    f. Can the text be read clearly from only one reading?
    g. Are all the titles appropriate for effective use by search engines?

5. Mechanics
    a. How functional is the website, and how fast does the site react and the pages load?
    b. Does the site's functionality, such as roll-down menus and mouse-over (if used), effectively support the use of the site?
    c. Are all the hyperlinks and buttons operating as per the stated requirements?
    d. Are there are error messages?

6. Concept
    a. What does the first page promise regarding what the rest of the website offers?
    b. What image of the organization does the site clearly project?
    c. What existing websites can be effectively compared with the one under consideration?
    d. What expectations will the website raise for its visitors?
    e. Can the rest of the site keep this promise effectively?

PROBLEMS

1. Discuss the important factors for considering usability during the software development process.
2. Discuss the steps for improving the usability of software products.
3. Describe the following software usability inspection methods:
    i. Cognitive walkthrough
    ii. Pluralistic walkthrough
4. Describe at least seven useful guidelines for conducting software usability testing.
5. Describe at least two widely used software testing methods.

6. Discuss at least six common web design–related errors.
7. Discuss at least four most important factors to be considered in web page design.
8. Discus at least three key factors to be considered in website design.
9. Describe the following two tools for evaluating web usability:
   i. NetRaker
   ii. Web SAT
10. Write an essay on software and web usability.

References

1. Keene, S. J., Software Reliability Concepts, *Proceedings of the Annual Reliability and Maintainability Tutorial Notes*, 1992, pp. 1–21.
2. Hopcroft, J. E., and Kraft, D. B., Sizing the U.S. Industry, *IEEE Spectrum*, December 1987, pp. 58–62.
3. Dhillon, B. S., *Engineering Usability: Fundamentals, Applications, Human Factors, and Human Error*, American Scientific Publishers, Los Angeles, 2004.
4. ISO/IEC 14598-1, *Software Product Evaluation: General Overview*, International Organization for Standardization (ISO), Geneva, Switzerland, 1999.
5. Powell, T., *Web Design: The Complete Reference*, Osborne McGraw-Hill, Berkley, CA, 2000.
6. Good, M., Software Usability Engineering, *Digital Technical Journal*, No. 6, February 1988, pp. 125–133.
7. Wixon, D., and Bramhall, M., How Operating Systems Are Used: A Comparison of VMS and UNIX, *Proceedings of the Human Factors Society 29th Annual Meeting*, 1985, pp. 245–249.
8. Whiteside, J. et al. How Do People Really Use Text Editors?, *SIGOA Newsletter*, No. 3, June 1982, pp. 29–40.
9. Good, J., Whiteside, D., Wixon, D., and Jones, S., Building a User-Derived Interface, *Communications of the ACM*, Vol. 27, October 1984, pp. 1032–1043.
10. Emerson, M., Porter, D., and Rudman, F., *Improving Software Usability: A Manager's Guide*, Report, Enervision Media, Inc., East Chatham, NY, 2002.
11. Fitzpatrick, R., *Strategies for Evaluating Software Usability, Report*, Department of Mathematics, Statistics, and Computer Science, Dublin Institute of Technology, Dublin, 2002.
12. Lee, S. H., Usability Testing for Developing Effective Interactive Multimedia Software: Concepts, *Dimensions, and Procedures, Educational Technology and Society*, Vol. 2, No. 2, 1999, pp. 100–113.
13. Dix, A., Finlay, J., Abowd, G., and Beale, R., *Human-Computer Interaction*, Prentice Hall, Inc., Hemel Hempstead, UK, 1998.
14. Avouris, N. M., *An Introduction to Software Usability*, Report, ECE Department, University of Patras, Rio Patras, Greece, 2002.
15. ISO 9241-11 (Draft International Standard), *Ergonomics Requirements for Office Work with Visual Display Terminals (VDT), Part III: Guidance on Usability*, International Organization for Standardization (ISO), Geneva, Switzerland, 1997.

16. Ferre, X., Juristo, N., Windl, H., and Constantine, L., Usability Basics for Software Developers, *IEEE Software*, January/February, 2001, pp. 22–30.
17. Preece, J., Rogers, Y., Sharp, H., Benyon, D., Holland, S., and Carey, T., *Human Computer Interaction*, Addison-Wesley, Inc., Reading, MA, 1994.
18. Preece, J., *Online Communities: Design Usability, Supporting Sociability*, John Wiley and Sons, New York, 2000.
19. Manning, H., McCarthy, J. C., and Sonza, R. K., *Why Most Web Sites Fail?*, White paper, Forrester Research, Cambridge, MA, September 1998.
20. Nielsen, J., PR on Websites: Increasing Usability, Alertbox, March 10, 2003. Available online at www.useit.com/alertbox/20030310.html.
21. Souza, R. K. et al. *The Best of Retail Site Design*, White paper, Forrester Research, Cambridge, MA, October 2000.
22. Chi, E. H., Improving Web Usability Through Visualization, *IEEE Internet Computing*, March/April 2002, pp. 64–71.
23. Trenner, L., and Bawa, J., editors, *The Politics of Usability: A Practical Guide to Designing Usable Systems in Industry*, Springer-Verlag, London, 1998.
24. Nielsen, J., Intranet Usability: The Trillion-Dollar Question, Alertbox, November 11, 2002. Available online at www.useit.com/alertbox/2002 1111.html.
25. Nielsen, J., Top Ten Mistakes in Web Design, Alertbox, May 1996. Available online at www.useit.com/alertbox/9605.html.
26. Nielsen, J., *Designing Web Usability: The Practice of Simplicity*, New Riders Publishing, Indianapolis, IN, 2000.
27. Brown, G. E., *Web Usability Guide, Report*, NEES Consortium, Inc., Richmond, CA, 2003. Available online at www.nees.org/info/contact-us.html.
28. Fleming, J., *Web Navigation: Designing the User Experience*, O'Reilly and Associates, Inc., Sebastopol, CA, 1998.
29. Williams, R., and Tollett, J., *The Non-Designer's Web Book: An Easy Guide to Creating, Designing and Posting Your Own Web site*, Peachpit Press, Berkeley, CA, 2000.
30. Price, J., Price, L., *Hot Text: Web Writing that Works*, New Riders Publishing, Indianapolis, IN, 2002.
31. Niederst, J., *Web Design in a Nutshell: A Desktop Quick Reference*, O'Reilly and Associates, Inc., Sebastopol, CA, 2001.

chapter twelve

Medical device usability and user errors

12.1   Introduction

Each year, a vast sum of money is spent on health care around the globe. For example, medical spending was around $938 billion in 1994 in the United States alone [1–3]. Today, there are approximately 6,500 hospitals and 700,000 doctors in the United States, and U.S. exporters control about half of the world's $71 billion medical device market [2,3].

Good user-interface design of medical device/equipment is extremely important for its safe and effective installation, operation, and maintenance [2,4,5]. Furthermore, various studies conducted over the years clearly indicate that poorly designed human-machine interfaces in medical devices/equipment significantly increase the risk of the occurrence of human error [2,5–7]. Further, medical device/equipment usability-related concerns, with regard to potential users such as physicians, nurses, patients, and professional caregivers, must be raised to the same level as traditional/conventional manufacturing, economic, and technological concerns during the design phase.

Each year, billions of dollars are spent to produce many types of engineering products, ranging from simple medical devices to highly sophisticated computer systems. Although most of these products are easy to use, some may need a very high degree of skill, and the user expertise in the necessary areas may vary considerably. Therefore, during the design phase of a complex engineering product, human factors normally are given considerable attention. The user/operator is studied at length, and the interface design is made user-friendly to minimize the occurrence of user/operator error. This chapter presents various aspects of medical device usability and user errors.

12.2   Medical device use description, users, use environments, and user interfaces

To understand a medical device's use completely and accurately, the appropriate description of its intended use is essential, including information on items such as the following [2,8]:

- User population characteristics—that is, those characteristics that can affect device use either directly or indirectly

- Use environments
- User interface design or preliminary design
- User-related needs for effective and safe use of the device and how the device meets them
- Device operation
- Normal use scenarios—that is, scenarios describing how the device will actually be utilized

A medical device may be fairly easy to use for one group of users but quite difficult for another. Thus, there is a definite need for devices that can be used safely and effectively by as many types of users as possible. Therefore, to meet this user need, it is essential to fully understand the limitations and abilities of all types of potential device users. Some examples of potential device users are professional health-care providers, patients, and young and elderly individuals. Factors such as fatigue, medication, and stress can affect the ability levels of device users, directly or indirectly.

Nonetheless, some of the important characteristics of the potential user population that should be taken into consideration carefully during medical device design are presented here [8]:

- Physical size, strength, memory, and cognitive ability
- Degree of knowledge about device operation and the associated medical condition
- Ability for adopting to adverse circumstances properly
- Normal health and mental state (e.g., relaxed, tired, and stressed) when using the device under consideration
- Earlier experience with similar user interfaces or devices
- Motivation and coordination (i.e., manual dexterity)
- Sensory-related capabilities (i.e., hearing, vision, and touch)

The use environments of medical devices can vary quite significantly and can have major impacts on their usability. Thus, with regard to the users of devices, factors such as the following must be considered carefully [1,2]:

- *Physical workload*: This is associated with the use of the device and it adds to the user stress. Under high stress, the device users are distracted and have less time for making decisions, considering multiple device outputs.
- *Motion and vibration*: This can significantly affect user ability to read displayed information effectively, to conduct fine physical manipulations such as typing on the keyboard portion of a medical device.

- *Mental workload*: This is the degree of thinking and concentration an individual exerts while using a medical device. When the mental workload imposed on users by the environment exceeds their abilities, the chances of improper use of devices increase dramatically. An example of this scenario could be an operating room having too many alarms on different devices, thus making it quite difficult for an anaesthetist to highlight the source of any single alarm.
- *Noise and light*: The effectiveness of auditory and visual displays (e.g., auditory alarms, lighted indicators, and other signals) can be very limited by the use environments if they are designed improperly. For example, in highly noisy environments, device users may not hear alarms if they are not sufficiently loud or distinctive. Nonetheless, some of the important considerations for displays, including visual alarm indicators, are ambient light levels, viewing angles, and existence of other devices in the use environment.

User interfaces of well-designed medical devices will facilitate appropriate actions and discourage or prevent the occurrence of hazardous actions. The user interface comprises all the elements of a medical device with which users interact while using the device, preparing it for use (e.g., setup and calibration), or conducting maintenance-related activities. More specifically, the user interface includes all hardware-related features that control the device's operation.

In modern medical devices, the user interfaces generally are computer-based. In this case, interface characteristics include items such as the following [2,8]:

- Mouse
- Keyboards
- Screen elements
- Control and monitoring screens
- Navigation logic
- Data entry requirements
- The manner in which data are presented and organized
- Alerting mechanisms
- Prompts
- Help functions

Finally, note that items such as operating instructions, training materials, packaging, and labeling are also considered part of the user interface, and thus they need careful consideration with regard to their effective usability.

12.3   General approach for developing effective user interfaces of medical devices

The general approach for developing effective user interfaces of medical devices is basically composed of six steps [2,3,9]:

1. *Step 1*: Define all project goals and system functionality.
2. *Step 2*: Conduct analysis of user capabilities, tasks, and work environments (note that this is also concerned with allocating tasks between human beings and the system).
3. *Step 3*: Document all user needs and requirements (note that this is also concerned with developing design prototypes and usability goals).
4. *Step 4*: Conduct usability testing (note that this is also concerned with evaluating the results of usability testing against performance goals and objectives, and a return to Step 3 is made as necessary).
5. *Step 5*: Develop appropriate design specifications for the device's user interface.
6. *Step 6*: Evaluate device interface designs during their field use (note that a return to step 3 may be needed as appropriate).

12.4   Useful guidelines for making medical device interfaces more user-friendly

Past experiences over the years clearly indicate that medical devices such as patient monitors, blood chemistry analyzers, kidney dialysis machines, infusion pumps, and ventilators often have various superficial user interface design–related problems. These problems can negatively affect a device's usability and appeal, directly or indirectly. However, such problems are relatively easy to remedy, and the following guidelines are considered quite useful to address them [2,10]:

- *Guideline I: Effective visual balance*—Generally, this is created about the vertical axis by arranging all involved visual elements on either side of an assumed axis. Each side contains roughly the same amount of content as empty space. There are various methods that can be employed for evaluating the balance of a composition. In turn, all perceived imbalances can be rectified through a number of means, including reorganizing information, adjusting the gaps between labels and field, relocating elements to other screens, or popping up elements only upon request.
- *Guideline II: Limit the usage of colors as much as possible*—This involves limiting the color palette of a device user interface, and two useful considerations in this area are presented here:

- Ensure that the selection of colors is clearly consistent with all involved medical conventions. For example, red is widely employed to symbolize alarm-associated information or to communicate arterial blood pressure values.
- Keep the colors of the background and major on-screen components somewhere between three and five, including shades of gray.
- *Guideline III: Use simple and straightforward language as much as possible*— Often medical device user interfaces are characterized by overly complex phrases and words that create various types of usability-related problems. These can easily be reduced, or eliminated altogether, by the use of simple and straightforward language. In this regard, there are many corrective measures, including using consistent syntax, writing shorter sentences, breaking cumbersome procedures into a number of ordered steps, and creating meaningful subheadings and headings.
- *Guideline IV: Reduce screen density as much as possible*—This is concerned with lowering the overstuffing of medical device displays with controls and information. The empty space generated by such reduction is very useful in a user interface because it helps to separate information into related groups and provides a resting place for users' eyes. Otherwise, overly dense-looking device user interfaces can be very intimidating to medical professionals such as nurses, technicians, and physicians, making it very difficult for these personnel to retrieve desired information at a glance. Nonetheless, actions such as those presented here can be very helpful to eliminate extraneous information on medical device displays:
  - Present secondary information on demand through appropriate pop-ups or relocate it, if possible, to other screens in a clear way.
  - Reduce text size as much as possible by stating things in a simpler manner.
  - Use simplified graphics.
  - Aim to utilize empty space rather than lines for separating content.
  - Decrease the size of graphics related to identity (i.e., brand names and logos) as much as possible.
- *Guideline V: Simplify typography as much as possible*—Efficient medical device user interfaces are based on typographical rules that help to make screen content easy and straightforward to read, and direct users effectively toward the important information first. Normally, this is achieved by having a single font and a few character sizes.

  Another approach utilized for simplifying typography is by eliminating or lowering excessive highlighting such as underlining, bolding, and italicizing. All in all, generally a single approach such as italicizing or bolding is quite sufficient to highlight important information effectively when employed in conjunction with various sizes of font and extra line spaces.

- *Guideline VI: Harmonize and refine icons*—This is a very important guideline, and some of the actions that can be taken to give the icons a family resemblance and maximize icon understanding are as follows [2,10]:
  - Simplify icon elements for eliminating unnecessary and confusing details.
  - Conduct user testing for ensuring that no two icons are so similar that they create confusion.
  - Make all icons that have similar purpose of the same overall size.
  - Develop a limited set of icon elements that clearly denote nouns only.
  - Reinforce all icons with text labels.
  - Use the same style for all similar-purpose icons.
- *Guideline VII: Ascribe to a grid*—Past experiences over the years clearly indicate that most screens normally function and look better when their all screen elements are aligned and they serve a utilitarian objective. Payoffs in terms of visual appeal and perceived simplicity are accrued by fitting all on-screen elements into a grid. Moreover, grid-based screens are normally easier to implement in computer code because of the visual elements' predictability. The following guidelines are quite useful with regard to ascribing to a grid:
  - Aim to keep all on-screen elements at fixed distances from the grid lines, as this can provide visual appeal.
  - Start by defining the screen's dominant elements and approximate space-related needs when developing a grid structure.
- *Guideline VIII: Provide appropriate navigation options and cues*—A navigator can sometimes become lost in a medical device user interface when going from one place to another. It is very much like the situation when a person is traveling in an unfamiliar city with no road signs or signs in a language that the person does not understand. More specifically, the problem occurs from device users not clearly knowing where they are in the user-interface structure.

  Actions such as those presented here are considered useful with regard to providing navigation options and cues:
  - Group together all navigation options and controls, such as "Help," "Go to main menu," and "Go back," in a single and consistent location. This will allow potential users to return easily to a previous screen or undo an action without any fear of getting lost.
  - Place effective titles on all screens and subcomponents by means of a header (i.e., a contrasting horizontal bar that effectively incorporates text).
  - Number the pages of electronic documents as you would for printed materials.

- *Guideline IX: Eradicate design inconsistencies as much as possible*—These inconsistencies are very disruptive to usability and user-interface appeal, and for some medical devices, they also may compromise safety. These design inconsistencies can be prevented by developing and maintaining a style guide that subsequently can be integrated into appropriate final design specifications.
- *Guideline X: Use hierarchical labels as much as possible*—The use of redundant labels leads to congested screens that generally take a long time to scan. Furthermore, excessive visual congestion generally makes it cumbersome for device users to pick out even the most salient details. Hierarchical labeling is very useful for saving space and speed scanning by displaying items such as heart rate, respiratory rate, and arterial blood pressure in a more efficient manner.

12.5   Designing medical devices for old users

The population of older people is growing at a significant rate in the United States. For example, it is forecast that by the year 2020, about 20% of Americans will be over 65 years of age [2,11,12]. This fact simply means that there is a definite need for designing medical devices for use by older people. This requires careful consideration during the device design to factors such as those shown in Figure 12.1 [2,11–15].

Elderly people are subject to various limitations; two common ones are impaired vision and hearing. Some elderly persons not only lose their ability to see well in low light, but also their ability to differentiate light intensities and colors. This means that the device designers must pay close attention to using somewhat oversized fonts for readouts, displays, and labels on devices designed to be used by older people. The decline in an individual's hearing ability is normally a function of age. Furthermore, a person's hearing tends to decrease with the increase in sound frequency, and females and males suffer greater hearing loss at frequencies in the 550–1000-Hz range and 3000–6000-Hz range, respectively, as they age. Therefore, designers must carefully consider factors such as these in medical devices to be used by elderly persons.

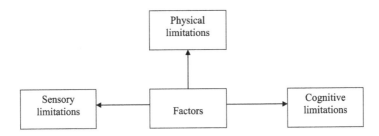

Figure 12.1   Factors to be considered in designing medical devices for older users.

Past experiences over the years clearly indicate that the cognition abilities of older individuals may vary considerably. More specifically, many of these persons may remain mentally sharp, but others may experience attention deficits often referred to as *cognitive rigidity*, a condition that makes it cumbersome to learn new procedures and methods. In this regard, it is highly recommended to decrease the number of steps in a specified procedure concerning a medical device in order to improve its usability effectiveness among elderly persons [2,11]. Another useful recommendation to facilitate user-related tasks for elderly individuals is to build redundant cues into medical device design [2,16].

Elderly users of medical devices are also subject to various physical limitations. For example, a significant number of people generally lose around 10%–20% of their strength by the time they reach 60–70 years of age. Moreover, the mobility of these personnel may be handicapped significantly by joint-related problems and associated pain. In order to overcome physical limitations such as these, designers should incorporate appropriate controls with large-diameter knobs into medical devices so that rotation needs a lesser degree of fine motor control, as well as textured knob surfaces that require less pinching strength to counter the issue of finger slippage, among other features.

12.6   Medical devices with high incidence of user/human error

In the area of health care, many types of medical devices are used, and a good number of these are prone to user/human error. In 1991, the U.S. Food and Drug Administration (FDA) conducted a study of device-related data collected over many years and highlighted the 20 medical devices most prone to user/human error [2,11,17]:

1. Glucose meter (highest error-prone)
2. Balloon catheter
3. Orthodontic bracket aligner
4. Administration kit for peritonea dialysis
5. Permanent pacemaker electrode
6. Implantable spinal cord stimulator
7. Intravascular catheter
8. Infusion pump
9. Urological catheter
10. Electrosurgical cutting and coagulation device
11. Nonpowered suction apparatus
12. Mechanical/hydraulic impotance device
13. Implantable pacemaker

14. Peritoneal dialysate delivery system
15. Catheter introducer
16. Catheter guide wire
17. Transluminal coronary angioplasty catheter
18. Low-energy defibrillator (external)
19. Continuous ventilator (respirator)
20. Contact lens cleaning and disinfecting solution (lowest error-prone)

Finally, the FDA study clearly indicates that errors in using medical devices, directly or indirectly, cause an average of at least three deaths or serious injuries in the United States each day [11].

12.7   Facts, figures, and examples on user/operator-related errors

Some of the important statistics and examples directly or indirectly concerned with user/operator error are as follows:

- As per the Center for Devices and Radiological Health (CDRH) of the FDA, about 60% of the deaths or serious injuries associated with medical devices were due to user error [2,18].
- A study reported that operator/user error accounts for over 50% of all technical-related medical equipment problems [2,19].
- Human error contributes to or causes up to 90% of accidents, both generally and in medical devices [20–22].
- An infant patient became hypoxic during treatment with oxygen because the attending physician set the flow control knob between 1 and 2 l/min without realizing that the scale numbers denoted discrete instead of continuous settings [23].
- A patient at a health-care facility was injured seriously because an attending nurse mistakenly read the number 7 as 1 [23].
- A study of air traffic control system-related errors reported that over 90% of these errors were due to operators [24].
- A study revealed that approximately 41% of the 631 accidents involving hot-water boilers in the United States in 2001 were due to either operator/user error or poor maintenance [2,25].
- A patient died because an infusion pump was mistakenly set to deliver 210 cc/h of heparin instead of the ordered 21 cc/h [26].
- A fatal radiation-overdose accident involving the Therac radiation therapy device was due to an operator/user error [27].
- In 2001, 1,091 accidents involving steam-heated boilers occurred in the United States. Operator/user error or poor maintenance was responsible for approximately 37% of these accidents [25].

12.8 Operator/user error causes and classifications of user errors in human-computer interactive tasks

There are many causes for operator/user errors. The main ones are as follows [2,28]:

- Poorly designed equipment/device
- Departure from following the correct operating procedures
- Operator/user carelessness
- Poor environment
- Task complexity and overload conditions
- Lack of appropriate procedures
- Lack of operator interest
- Poor personnel selection and training

User errors occurring in human-computer interactive tasks may be classified under the following broad categories [2,29]:

- *Classification I: Omission errors*—These include memory and attention errors. With regard to computer interactive tasks, two typical examples of omission errors are as follows:
  - *Example 1*: Quitting the application without saving the work
  - *Example 2*: Clicking on the paste text icon without copying the text
- *Classification II: Commission errors*—There are four types of commission errors: sequence errors, selection errors, time errors, and qualitative errors. Sequence errors occur when users conduct tasks in the incorrect order. For example, users of the "merge" feature encountered problems comprehending steps in the procedure, and so they conducted them in an incorrect order.

  Selection errors occur when users select an incorrect option to perform a task. Three examples of selection errors are clicking the wrong icon, selecting the wrong menu option, and choosing the incorrect option in a dialog box. Finally, time errors and qualitative errors are considered self-explanatory.

12.9 Common medical device/equipment operator/ user errors

Each year, billions of dollars are spent to produce various types of medical devices and equipment throughout the world [30]. These devices/ equipment range from the relatively simple, such as catheters and syringes, to the highly technologically complex and sophisticated, such as computer-controlled diagnostic equipment, and all of these are subject to user-related

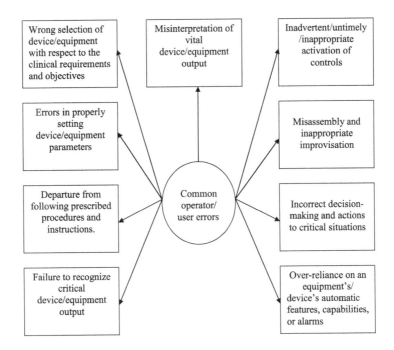

Figure 12.2   Common medical device/equipment operator/user errors.

errors. Users of such devices and equipment include nurses, physicians, patients, and support personnel.

Over the years, various types of studies have been carried out to investigate the occurrence of errors in the use of medical devices and equipment [7]. Some of the common operator/user errors found in these studies are shown in Figure 12.2 [2,31].

12.10   Methods for performing user error–related analysis

Over the years, many methods and techniques have been developed in human factors, human reliability, and reliability areas for performing various types of analysis. Some of these methods also can be used to perform user error analysis. This chapter presents four of these methods: probability tree analysis, failure modes and effect analysis (FMEA), the Markov method, and fault tree analysis (FTA).

12.10.1   Probability tree analysis

This method was developed for performing human reliability analysis in the nuclear power industrial sector [32]. It is very useful for performing task analysis by diagrammatically representing critical human actions, as well as other events associated with the system under consideration. Diagrammatic representations lead to a probability tree structure. The method is described in more detail in Chapter 4.

12.10.2   Failure modes and effect analysis (FMEA)

FMEA is a widely used method in the industrial sector for performing reliability and safety studies of engineering systems. The method was developed in the early 1950s for conducting safety analysis of flight control systems [33]. When the method is extended to categorize each failure effect according to its severity, it is known as *failure mode effects and criticality analysis (FMECA)* [34]. All in all, FMEA can also be used for performing user/operator error analysis and is described in more detail in Chapter 4.

12.10.3   Markov method

The Markov method is widely used in the industrial sector for performing various types of reliability and availability analysis, particularly when the reparability of items is involved. The method is named after the Russian mathematician Andrei Andreyevich Markov (1856–1922) and is described in detail in Chapter 4. The applicability of the method to user error–related problems is demonstrated through the two mathematical models presented next.

12.10.3.1   Model I

Model I represents a user or operator carrying out a time-continuous task. The person can make an error in carrying out this task. The model state space diagram is shown in Figure 12.3, and the numerals in circles denote corresponding states. For a given constant user/operator error rate, the Markov method can be utilized for predicting operator/user reliability at time t, user/operator probability of committing an error at time t, and user/operator mean time to error.

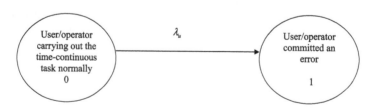

Figure 12.3 State space diagram for a user or operator carrying out a time-continuous task.

The following symbols are associated with this model:

j is the user/operator state; $j = 0$ means that the user/operator is carrying out the time-continuous task normally, $j = 1$ means that the user/operator committed an error.

$P_j(t)$ is the probability that the user/operator is in state j at time t; for $j = 1, 2$.

λ_u is the user/operator constant error rate.

s is the Laplace transform variable.

By using the Markov method, we write the following differential equations for the state space diagram in Figure 12.3:

$$\frac{dP_0(t)}{dt} + \lambda_u P_0(t) = 0 \tag{12.1}$$

$$\frac{dP_1(t)}{dt} - \lambda_u P_1(t) = 0. \tag{12.2}$$

At time $t = 0$, $P_0(0) = 1$ and $P_1(0) = 0$.

Solving Equations 12.1 and 12.2 by using the Laplace transforms, we obtain

$$P_0(s) = \frac{1}{s + \lambda_u} \tag{12.3}$$

$$P_1(s) = \frac{\lambda_u}{s(s + \lambda_u)}, \tag{12.4}$$

where $P_0(s)$ is the Laplace transform of the probability that the user or operator is in state 0 and $P_1(s)$ is the Laplace transform of the probability that the user or operator is in state 1.

After taking the inverse Laplace transform of Equations 12.3 and 12.4, we get

$$P_0(t) = e^{-\lambda_u t} \tag{12.5}$$

$$P_1 = 1 - e^{-\lambda_u t}. \tag{12.6}$$

Equation 12.5 is the user/operator reliability at time t. By integrating Equation 12.5 over the time interval $[0, \infty]$, we obtain

$$\mathrm{MTTUE} = \int_0^\infty e^{-\lambda_u t} dt \tag{12.7}$$

$$= \frac{1}{\lambda_u},$$

where MTTUE is the mean time to user/operator error.

EXAMPLE 12.1

Assume that a user or operator is carrying out a certain task, and his or her error rate is 0.0001 errors/h. Calculate the following:

- Probability of the user or operator committing an error during a 100-h mission.
- Mean time to user/operator error.

By inserting the given data values into Equations 12.6 and 12.7, we obtain

$$P_1(100) = 1 - e^{-(0.0001)(100)}$$
$$= 0.0099$$

and

$$\text{MTTUE} = \frac{1}{(0.0001)}$$
$$= 10,000 \text{ h.}$$

Thus, the probability of the user or operator committing an error during a 100-h mission and the mean time to user/operator error are 0.0099 and 10,000 h, respectively.

12.10.3.2   *Model II*

Model II is basically the same as Model I, but instead of the user or operator carrying out his or her task in a constant environment, he or she carries out tasks in a fluctuating environment (i.e., normal and abnormal or normal and stressful) [35]. The user or operator can commit an error in either a normal or an abnormal environment. Moreover, the error rate of the user or operator may increase significantly in a stressful environment.

The model state space diagram is shown in Figure 12.4; the numerals in circles and boxes denote corresponding states.

The following symbols are associated with this model:

j is the user/operator state; $j = 1$ means that the user or operator is carrying out the task correctly in a normal environment, $j = 1$ means that the user or operator committed an error in a normal environment, $j = 2$ means that the user or operator is carrying out the task correctly in an abnormal environment, $j = 3$ means that the user or operator committed an error in an abnormal environment.
$P_j(t)$ is the probability that the user or operator is in state j at time t; $j = 0,1,2,3$.
λ_n is the user/operator constant error rate in a normal environment.

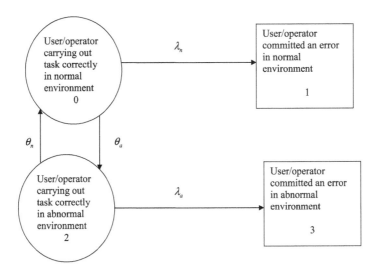

Figure 12.4 State space diagram for the user or operator carrying out tasks under fluctuating environment.

λ_a is the user/operator constant error rate in an abnormal environment.
θ_n is the constant transition rate from an abnormal or stressful environment to a normal environment.
θ_a is the constant transition rate from a normal environment to an abnormal or stressful environment.

By using the Markov method, we write the following set of differential equations for the state space diagram in Figure 12.4:

$$\frac{dP_0(t)}{dt} + (\lambda_n + \lambda_a)P_0(t) = \theta_n P_2(t) \tag{12.8}$$

$$\frac{dP_2(t)}{dt} + (\lambda_a + \theta_n)P_2(t) = \theta_a P_0(t) \tag{12.9}$$

$$\frac{dP_1(t)}{dt} = \lambda_n P_0(t) \tag{12.10}$$

$$\frac{dP_3(t)}{dt} = \lambda_a P_2(t). \tag{12.11}$$

At time $t = 0$, $P_0(0) = 1$, and $P_1(0) = P_2(0) = P_3(0) = 0$.
By solving Equations 12.8 through 12.11, we obtain

$$P_0(t) = (N_1 - N_2)^{-1}[(N_2 + \lambda_a + \theta_n)e^{N_2 t} - (N_1 + \lambda_a + \theta_n)e^{N_1 t}], \tag{12.12}$$

where

$$N_1 = \frac{-k_1 + \left(k_1^2 - 4k_2\right)^{1/2}}{2}$$

$$N_2 = \frac{-k_1 - \left(k_1^2 - 4k_2\right)^{1/2}}{2}$$

$$k_1 = \lambda_n + \lambda_a + \theta_n + \theta_a$$

$$k_2 = \lambda_n(\lambda_a + \theta_n) + \theta_a\lambda_a$$

$$P_1(t) = k_4 + k_5 e^{N_2 t} - k_6 e^{N_1 t}, \tag{12.13}$$

where

$$k_3 = \frac{1}{N_1 - N_2}$$

$$k_4 = \frac{\lambda_n(\lambda_a + \theta_n)}{(N_1 + N_2)}$$

$$k_5 = k_3(\lambda_n + k_4 N_1)$$

$$k_6 = k_3(\lambda_n + k_4 N_2)$$

$$P_2(t) = \theta_a k_3 \left(e^{N_2 t} - e^{N_1 t}\right) \tag{12.14}$$

$$P_3(t) = k_7 \left[(1 + k_3)\left(N_1 e^{N_2 t} - N_2 e^{N_1 t}\right)\right], \tag{12.15}$$

where

$$K_7 = \frac{(\lambda_a \theta_a)}{N_1 N_2}.$$

The user/operator reliability in an alternating environment is given by

$$R_u(t) = P_0(t) + P_2(t), \tag{12.16}$$

where $R_u(t)$ is the user/operator reliability at time t in an alternating environment.

The mean time to user/operator is expressed by

$$
\begin{aligned}
\mathrm{MTTUE}_f &= \int_0^\infty R_u(t)dt \\
&= \frac{(\lambda_a + \theta_a + \theta_n)}{k_2},
\end{aligned}
\tag{12.17}
$$

where MTTUE_f is the mean time to user/operator error in fluctuating environment.

EXAMPLE 12.2

Assume that a user/operator conducts his or her time-continuous tasks in an alternating (i.e., normal and abnormal) environment and his or her constant error rates are 0.0002 errors/h and 0.0006 errors/h, respectively. The constant transition rates from normal to abnormal and vice versa are 0.0008/h and 0.0005/h, respectively. Calculate the mean time to user/operator error.

By inserting the specified data values into Equation 12.17, we get

$$
\mathrm{MTTUE}_f = \frac{(0.0006 + 0.0008 + 0.0005)}{(0.0002)(0.0006 + 0.0005) + (0.0008)(0.0006)}
$$

$$
= 2{,}714.28 \text{ h}
$$

Thus, the mean time to user/operator error is 2,714.28 h.

12.10.4 Fault tree analysis (FTA)

FTA is one of the most widely used methods in the industrial sector for performing reliability and safety analysis of engineering systems. The method was developed in the early 1960s for doing safety analysis of the Minuteman Launch Control system [2,36]. FTA is described in detail in Chapter 4. It can be used for performing user/operator error analysis. The following example demonstrates how FTA can be used to perform user error analysis.

EXAMPLE 12.3

Assume that user error A occurs if error X, Y, or Z occurs. In turn, error X occurs if errors x_1 and x_2 occur, and error Z occurs if error z_1 or z_2 occurs. Develop a fault tree for the top event "user error A occurrence" by using the fault tree symbols given in Chapter 4.

A fault tree for the example using the fault tree symbols given in Chapter 4 is shown in Figure 12.5.

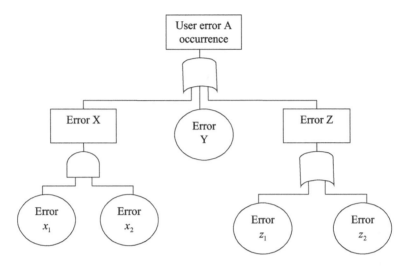

Figure 12.5   A fault tree for the top event: user error A occurrence.

EXAMPLE 12.4

Assume that in Example 12.3, the occurrence probabilities of error Y, x_1, x_2, z_1, and z_2 are 0.01, 0.02, 0.03, 0.04, and 0.05, respectively. Calculate the probability of the occurrence of user error A using the Figure 12.5 fault tree.

By using the given data values and the equation given in Chapter 4 for calculating the AND gate output event occurrence probability, we get the error X occurrence probability as follows:

$$P(X) = P(x_1)P(x_2)$$
$$= (0.02)(0.03)$$
$$= 0.0006,$$

where $P(x_1)$ is the occurrence probability of error x_1 and $P(x_2)$ is the occurrence probability of error x_2.

By using the given data values and the equation given in Chapter 4 for calculating the OR gate output event occurrence probability, we get the error Z occurrence probability as follows:

$$P(Z) = P(z_1) + P(z_2) - P(z_1)P(z_2)$$
$$= 0.04 + 0.05 - (0.04)(0.05)$$
$$= 0.088$$

By using the given data value, the calculated values, and the equation given in Chapter 4 for calculating the OR gate output event occurrence probability, we get the user error A occurrence

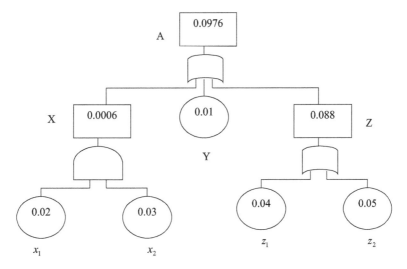

Figure 12.6   A fault tree with error occurrence probability values (plain and sub-script letters in the diagram denote errors).

probability as follows:

$$P(A) = 1 - (1 - P(X))(1 - P(Y))(1 - P(Z))$$
$$= 1 - (1 - 0.0006)(1 - 0.01)(1 - 0.088)$$
$$= 0.0976.$$

Thus, the probability of the user error A occurrence is 0.0976. Figure 12.6 shows the Figure 12.5 fault tree with given and calculated error occurrence probability values.

PROBLEMS

1. Discuss at least four factors regarding users of medical devices that can have major impacts on the usability of these devices.
2. Describe the general approach for developing effective user interfaces of medical devices.
3. Discuss at least eight guidelines for making medical device interfaces more user-friendly.
4. Discuss the factors to be considered in designing medical devices for elderly users.
5. What are the medical devices with high incidence of user error?
6. List at least eight main causes for the occurrence of operator or user errors.

7. List at least eight common medical device/equipment operator or user errors.
8. Discuss at least four methods that can be used to perform user error analysis.
9. Assume that a user or operator performs his or her time-continuous tasks in a fluctuating (i.e., normal and abnormal) environment and his or her constant error rates are 0.001 errors/h and 0.004 errors/h, respectively. The constant transition rates from normal to abnormal and vice versa are 0.007/h and 0.003/h, respectively. Calculate the mean time to user/operator error.
10. Assume that user error N occurs if error A, B, C, or D occurs. In turn, error A occurs if errors a_1, a_2, and a_3 occur and error D occurs, if error d_1, d_2, or d_3 occurs. Develop a fault tree for the top event "User error N occurrence" by using the fault tree symbols given in Chapter 4.

References

1. Tully, S., Why Drug Prices Will Go Lower. The industry's days of gargantuan earnings are gone forever. Why? Managed-care outfits are demanding price breaks, and generics are proving tough competitors. *Fortune Magazine* (USA), May 3, 1993, Vol. 127, No. 9. https://money.cnn.com/magazines/fortune/fortune_archive/1993/05/03/77806/index.htm
2. Dhillon, B. S., *Engineering Usability: Fundamentals, Applications, Human Factors, and Human Error*, American Scientific Publishers, Stevenson Ranch, CA, 2004.
3. Information Infrastructure for Healthcare, *Advanced Technology Program, National Institute of Standards and Technology (NIST)*, U.S. Department of Commerce, Washington, DC, 2003. Available online at http://www.atp.nist.gov/www/press/iih9703.htm
4. Sawyer, D., *Do It By Design: An Introduction to Human Factors in Medical Devices*, Center for Device and Radiological Health, U.S. Food and Drug Administration, Washington, DC, 1997.
5. Garmer, K., Liljegren, E., Osvalder, A. L., and Dahlman, S., Arguing for the Need of Triangulation and Iteration When Designing Medical Equipment, *Journal of Clinical Monitoring and Computing*, Vol. 17, 2002, pp. 105–114.
6. Obradovich, J. H., and Woods, D. D., Users as Designers: How People Cope with Poor HCI Design in Computer-Based Medical Devices, *Human Factors*, Vol. 38, 1996, pp. 40–46.
7. Hayman, W. A., Errors in the Use of Medical Equipment, in *Human Error in Medicine*, edited by M. S. Bogner, Lawrence Erlbaum Associates, New York, 1994, pp. 327–347.
8. Medical Device Use-Safety, *Incorporating Human Factors Engineering into Risk Management, Draft Guidance Document*, Center for Devices and Radiological Health, U.S. Food and Drug Administration, Washington, DC, 2000.

9. Salvemini, A. J., Challenge for User-Interface Designers of Telemedicine Systems, *Telemedicine Journal*, Vol. 5, No. 2, 1999, pp. 10–15.
10. Wiklund, M. E., Making Medical Device Interfaces More User Friendly, *Medical Device & Diagnostic Industry (MDDI) Magazine*, May 1998, pp. 177–184.
11. Wiklund, M. E., *Medical Device and Equipment Design: Usability Engineering and Ergonomics*, Interpharm Press, Inc., Buffalo Grove, IL, 1995.
12. Czaja, S., Special Issue Preface, *Human Factors*, Vol. 32, No. 5, 1990, p. 505.
13. Czaja, S., Clark, M., and Weber, R., Computer Communication Among Older Adults, Proceedings of the Human Factors Society 34th Annual Meeting, 1990, pp. 304–309.
14. CPSC Publication No. 702, *Product Safety and the Older Consumer: What Manufacturers/Designers Need to Consider*, Consumer Product Safety Commission (CPSC), Washington, DC, 1988.
15. Small, A., Design for Older People, in *Handbook of Human Factors*, edited by G. Salvendy, John Wiley & Sons, New York, 1987, pp. 125–140.
16. Koncelik, J., *Aging and the Product Environment*, Scientific and Academic Additions, Inc., Florence, KY, 1982.
17. Medical Device Reporting (MDR) System, *Center for Devices and Radiological Health (CDRH)*, U.S. Food and Drug Administration, Washington, DC, 1991.
18. Bogner, M. S., Medical Devices: A New Frontier for Human Factors, *CSERIAC Gateway*, Vol. IV, No. 1, 1993, pp. 12–14.
19. Dhillon, B. S., Reliability Technology in Health Care Systems, Proceedings of the IASTED International Symposium on Computers and Advanced Technology in Medicine, Health Care, and Bio-Engineering, 1990, pp. 84–87.
20. Bogner, M. S., Medical Devices and Human Error, in *Human Performance in Automated Systems: Current Research and Trends*, edited by M. Moulona and R. Parasuraman, Lawrence Erlbaum Associates, Hillsdale, NJ, 1994, pp. 64–67.
21. Maddox, M. E., Designing Medical Devices to Minimize Human Error, *Medical Device and Diagnostics Magazine*, Vol. 19, No. 5, 1997, pp. 166–180.
22. Nobel, J. L., Medical Device Failures and Adverse Effects, *Pediatric Emergency Care*, Vol. 7, 1991, pp. 120–123.
23. Sawyer, D., *Do It by Design: Introduction to Human Factors in Medical Devices*, Center for Devices and Radiological Health (CDRH), Food and Drug Administration (FDA), Washington, DC, 1996.
24. Kenney, G. C., Spahn, M. J., and Amato, R. A., *The Human Element in Air Traffic Control: Observations and Analysis of Controllers and Supervisors in Providing Air Traffic Control Services*, Report No. MTR-7655, METREK Div., MITRE Corp., December 1977.
25. Report No. AC42, 2002. Available from the National Board of Boiler and Pressure Vessel Inspectors, 1055 Crupper Avenue, Columbus, OH.
26. Brueley, M. E., Ergonomics and Error: Who is Responsible? Proceedings of the First Symposium on Human Factors in Medical Devices, 1989, pp. 6–10.

27. Casey, S., *Set Phasers on Stun: And Other True Tales of Design Technology and Human Error*, Aegean Inc., Santa Barbara, CA, 1993.
28. Meister, D., *Human Factors: Theory and Practice*, John Wiley & Sons, New York, 1976.
29. Lincoln, A., Human Factors in the Attainment of Reliability, *IRE Transactions on Reliability and Quality Control*, Vol. 11, 1962, pp. 97–103.
30. Murray, K., Canada's Medical Device Industry Faces Cost Pressures, Regulatory Reform, *Medical Device and Diagnostics Magazine*, Vol. 19, No. 8, 1997, pp. 30–39.
31. Hyman, W. A., Human Factors in Medical Devices, in *Encyclopaedia of Medical Devices and Instrumentation*, edited by J. G. Webster, Vol. 3, John Wiley & Sons, New York, 1988, pp. 1542–1553.
32. Swain, A. D., *A Method for Performing a Human-Factors Reliability Analysis*, Report No. SCR-685, Sandia Corporation, Albuquerque, New Mexico, August 1963.
33. Arnzen, H. E., *Failure Mode and Effect Analysis: A Power Engineering Tool for Component and System Optimization*, Report No. 347.40.00-K4-05 (C5776), GIDEP Operations Center, Corona, California, 1966.
34. Jordan, W. E., Failure Modes, Effects and Critical Analyses, Proceedings of the Annual Reliability and Maintainability Symposium, 1972, pp. 30–37.
35. Dhillon, B. S., Stochastic Models for Predicting Human Reliability, *Microelectronics and Reliablity*, Vol. 25, 1985, pp. 729–752.
36. Haasl, D. F., Advanced Concepts in Fault Tree Analysis, Proceedings of the System Safety Symposium, 1965, pp. 100–105. Available from the University of Washington Library, Seattle.

chapter thirteen

Mathematical models for engineering systems reliability analysis and usability assurance

13.1   Introduction

Mathematical models are widely used in engineering for studying many types of phenomena. In such models, the parts of an item or system are represented by idealized elements containing all the representative characteristics of the real-life parts/components and whose behavior is possible to be described by various equations. However, the degree of realism of a mathematical model depends very much on assumptions imposed upon it.

Over the years, many mathematical models and formulas have been developed to perform various types of studies in the areas of reliability and usability engineering. Although the effectiveness of these models can vary from one application to another, some of them can be used directly or indirectly for systems reliability analysis and usability assurance.

This chapter presents a number of mathematical models/formulas considered useful for systems reliability analysis and usability assurance.

13.2   Model I

Model I represents a system that can fail either due to a hardware failure or a user error. The system state space diagram is shown in Figure 13.1 [1,2]; the numerals in the circle and boxes denote system states.

The following assumptions are associated with this model:

- System hardware failures and user errors occur independently.
- System hardware failure and user error rates are constant.

The following symbols are associated with the diagram shown in Figure 13.1 and its associated equations:

$P_i(t)$ is the probability that the system is in state i at time t, for $i = 0$ (working normally), $i = 1$ (failed due to a hardware failure), $i = 2$ (failed due to a user error).

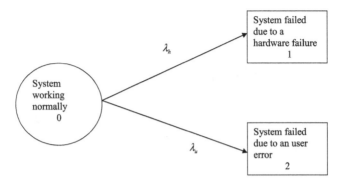

Figure 13.1 State space diagram for system failure due to hardware failure or user error.

λ_h is the system hardware failure rate.
λ_u is the system user error rate.

Using the Markov method described in Chapter 4, we write the following equations for the state-space diagram shown in Figure 13.1 [1,2]:

$$\frac{dP_0(t)}{dt} + (\lambda_u + \lambda_h)P_0(t) = 0 \tag{13.1}$$

$$\frac{dP_1(t)}{dt} - \lambda_h P_0(t) = 0 \tag{13.2}$$

$$\frac{dP_2(t)}{dt} - \lambda_u P_0(t) = 0. \tag{13.3}$$

At time $t = 0$, $P_0(0) = 1$, $P_1(0) = 0$, and $P_2(0) = 0$.
    By solving Equations 13.1 through 13.3, we get

$$P_0(t) = e^{-(\lambda_u + \lambda_h)t} \tag{13.4}$$

$$P_1(t) = \frac{\lambda_h}{\lambda_u + \lambda_h}(1 - e^{-(\lambda_u + \lambda_h)t}) \tag{13.5}$$

$$P_2(t) = \frac{\lambda_u}{\lambda_u + \lambda_h}(1 - e^{-(\lambda_u + \lambda_h)t}). \tag{13.6}$$

The system reliability is given by

$$R_s(t) = P_0(t)$$
$$= e^{-(\lambda_u + \lambda_h)t}, \tag{13.7}$$

where $R_s(t)$ is the system reliability at time t.

The system mean time to failure (SMTTF) is given by References 1,2, as follows:

$$\text{SMTTF} = \int_0^\infty R_s(t)dt$$

$$= \int_0^\infty e^{-(\lambda_u + \lambda_h)t} dt \tag{13.8}$$

$$= \frac{1}{\lambda_u + \lambda_h}.$$

EXAMPLE 13.1

Assume that a system can fail either due to a hardware failure or a user error, and its constant hardware failure and user error rates are 0.004 failures/h and 0.001 errors/h, respectively. Calculate the system reliability for a 100-h mission and mean time to failure (MTTF).

By substituting the given data values into Equation 13.7, we get

$$R_s(100) = e^{-(0.001+0.004)(100)}$$
$$= 0.6065.$$

Similarly, by substituting the given data values into Equation 13.8, we get

$$\text{SMTTF} = \frac{1}{(0.001 + 0.004)}$$
$$= 200 \text{ h}.$$

Thus, the system reliability and MTTF for the given data values are 0.6065 and 200 h, respectively.

13.3   *Model II*

Model II represents a system operating in fluctuating environments (i.e., normal and abnormal), such as stormy weather. The system state space diagram is shown in Figure 13.2 [1–3]; the numerals in circles and boxes denote the corresponding system states.

This model is subjected to the following assumptions:

- System failures and changes in environments occur independently.
- System failure rates are constant.
- The rates of changing environment from normal to abnormal and vice versa are constant.

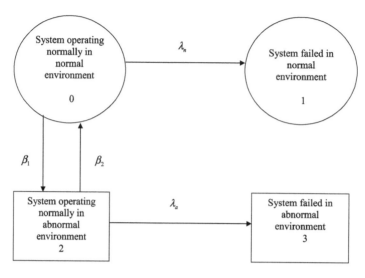

Figure 13.2 State space diagram for a system operating in fluctuating environments.

The following symbols are associated with this model:

$P_j(t)$ is the probability of the system being in state j at time t, for $j = 0,1,2,3$.

λ_n is the system constant failure rate in a normal environment.

λ_a is the system constant failure rate in an abnormal environment.

β_1 is the constant transition rate from a normal environment to an abnormal environment.

β_2 is the constant transition rate from an abnormal environment to a normal environment.

With the aid of the Markov method described in Chapter 4, we write the following set of differential equations for the state space diagram shown in Figure 13.2 [1–3]:

$$\frac{dP_0(t)}{dt} + (\lambda_n + \beta_1)P_0(t) = \beta_2 P_2(t) \tag{13.9}$$

$$\frac{dP_1(t)}{dt} = \lambda_n P_0(t) \tag{13.10}$$

$$\frac{dP_2(t)}{dt} + (\lambda_a + \beta_2)P_2(t) = \beta_1 P_0(t) \tag{13.11}$$

$$\frac{dP_3(t)}{dt} = \lambda_a P_2(t). \tag{13.12}$$

At time $t = 0$, $P_0(0) = 1$, $P_1(0) = P_2(0) = P_3(0) = 0$.

By solving Equations 13.9 through 13.12, we obtain

$$P_0(t) = (x_1 - x_2)^{-1}[(x_2 + \lambda_a + \beta_2)e^{x_2 t} - (x_1 + \lambda_a + \beta_2)e^{x_1 t}], \qquad (13.13)$$

where

$$x_1 = \frac{-b_1 + \sqrt{b_1^2 - 4b_2}}{2}$$

$$x_2 = \frac{-b_1 - \sqrt{b_1^2 - 4b_2}}{2}$$

$$b_1 = \lambda_n + \lambda_a + \beta_1 + \beta_2$$

$$b_2 = \lambda_n(\lambda_a + \beta_2) + \beta_1 \lambda_a$$

$$P_1(t) = b_4 + b_5 e^{x_2 t} - b_6 e^{x_1 t}, \qquad (13.14)$$

where

$$b_3 = \frac{1}{x_2 - x_1}$$

$$b_4 = \frac{\lambda_n(\lambda_a + \beta_2)}{x_1 x_2}$$

$$b_5 = b_3(\lambda_n + b_4 x_1)$$

$$b_6 = b_3(\lambda_n + b_4 x_2)$$

$$P_2(t) = \beta_1 b_3 (e^{x_2 t} - e^{x_1 t}) \qquad (13.15)$$

$$P_3(t) = b_7[(1 + b_3)(x_1 e^{x_2 t} - x_2 e^{x_1 t})], \qquad (13.16)$$

where

$$b_7 = \frac{\lambda_a \beta_1}{x_1 x_2}.$$

The system reliability in a fluctuating environment is expressed by

$$R_{sf}(t) = P_0(t) + P_2(t), \qquad (13.17)$$

where $R_{sf}(t)$ is the system reliability in a fluctuating environment at time t.

The mean time to system failure in a fluctuating environment is expressed by

$$\text{MTTSF}_f = \int_0^\infty R_{sf}(t)dt$$

$$= \frac{(\lambda_a + \beta_1 + \beta_2)}{b_2},$$

(13.18)

where MTTSF_f is the mean time to system failure in a fluctuating environment.

EXAMPLE 13.2

Assume that a system is operating in a fluctuating (i.e., normal and abnormal) environment, and its failure rates are 0.004 failures/h and 0.006 failures/h, respectively. The constant transition rates from normal to abnormal environment and from abnormal to normal environment are 0.008 per hour and 0.003 per hour, respectively. Calculate the MTTF of the system operating in a fluctuating environment.

By substituting the given data values into Equation 13.18, we obtain

$$\text{MTTSF}_f = \frac{(0.006 + 0.008 + 0.003)}{(0.004)(0.006 + 0.003) + (0.008)(0.006)}$$

$$= 202.38 \text{ hours}$$

Thus, the MTTF of the system operating in a fluctuating environment is 202.38 h.

13.4 Model III

Model III represents an engineering system that can be in any one of three states: system operating normally in the field, system failed in the field, and failed system in the workshop for repair. The failed engineering system is always taken to the workshop for repair, and after repair, it is put cback to its normal operating state. An example of such system is a motor vehicle.

The engineering system state space diagram is shown in Figure 13.3; the numerals in the boxes and circle denote system states. The following assumptions are associated with this model:

- System failure, towing, and repair rates are constant.
- The repaired system is as good as new.
- System failures occur independently.

The following symbols are associated with the state space diagram shown in Figure 13.3 and its associated equations:

$P_i(t)$ is the probability that the engineering system is in state i at time t, for $i = 0,1,2$.
λ_s is the system constant failure rate.
λ_{st} is the system towing rate to repair workshop from state 1.
γ is the system constant repair rate.

With the aid of the Markov method described in Chapter 4, we write the following set of differential equations for the state space diagram shown in Figure 13.3 [4]:

$$\frac{dP_0(t)}{dt} + \lambda_s P_0(t) = \gamma P_2(t) \tag{13.19}$$

$$\frac{dP_1(t)}{dt} + \lambda_{st} P_1(t) = \lambda_s P_0(t) \tag{13.20}$$

$$\frac{dP_2(t)}{dt} + \gamma P_2(t) = \lambda_{st} P_1(t). \tag{13.21}$$

At time $t = 0$, $P_0(0) = 1$, $P_1(0) = 0$, and $P_2(0) = 0$.

By solving Equations 13.19 through 13.21, we get the following steady-state probability equations [4]:

$$P_0 = \left[1 + \frac{\lambda_s}{\lambda_{st}} + \frac{\lambda_s}{\gamma} \right]^{-1} \tag{13.22}$$

$$P_1 = P_0 \left(\frac{\lambda_s}{\lambda_{st}} \right) \tag{13.23}$$

$$P_2 = P_0 \left(\frac{\lambda_s}{\gamma} \right), \tag{13.24}$$

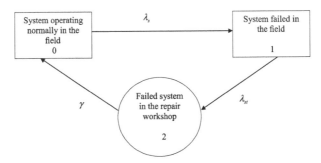

Figure 13.3   Engineering system state space diagram.

where P_0, P_1, and P_2 are the steady-state probabilities of the engineering system in states, 0, 1, and 2, respectively.

The engineering system's steady-state availability is given by

$$AV_s = P_0, \tag{13.25}$$

where AV_s is the engineering system's steady-state availability.

By setting $\gamma = 0$ in Equations 13.19 through 13.21 and then solving the resulting equations, we obtain

$$R_s(t) = P_0(t) = e^{-\lambda_s t}, \tag{13.26}$$

where $R_s(t)$ is the engineering system reliability at time t.

The engineering system's MTTF is given by

$$
\begin{aligned}
\text{MTTF}_s &= \int_0^\infty R_s(t)dt \\
&= \int_0^\infty e^{-\lambda_s t} dt \\
&= \frac{1}{\lambda_s},
\end{aligned}
\tag{13.27}
$$

where MTTF_s is the engineering system's MTTF.

EXAMPLE 13.3

Assume that the constant failure rate of an engineering system is 0.006 failures/h. Calculate the engineering system's MTTF and its reliability during a 10-h mission.

By inserting the specified data value into Equation 13.27, we get

$$
\begin{aligned}
\text{MTTF}_s &= \frac{1}{0.006} \\
&= 166.6 \text{ h}
\end{aligned}
$$

Using the specified data values in Equation 13.26, we obtain

$$
\begin{aligned}
R_s(10) &= e^{-(0.006)(10)} \\
&= 0.9417.
\end{aligned}
$$

Thus, the engineering system's MTTF and reliability are 166.6 h and 0.9417, respectively.

13.5 Model IV

Model IV represents a system that can fail safely or unsafely due to either a hardware failure or a user error. The failed system is repaired, and the state space diagram of the system is shown in Figure 13.4; the numerals in boxes and circle in Figure 13.4 denote corresponding system states.

The following assumptions are associated with this model:

- The system failure and repair rates are constant.
- The repaired system is as good as new.
- System failures occur independently.

The following symbols are associated with the diagram shown in Figure 13.4 and its associated equations:

$P_i(t)$ is the probability that the system is in state i at time t, for $i = 0,1,2$.
λ_s is the system constant safe failure rate.
λ_u is the system constant unsafe failure rate.
μ_s is the failed system constant repair rate from the safe failed state 1.
μ_u is the failed system constant repair rate from the unsafe failed state 2.

With the aid of the Markov method described in Chapter 4, we write the following set of differential equations for the state space diagram

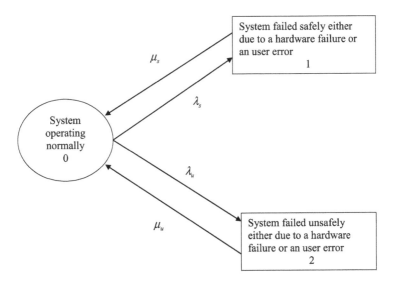

Figure 13.4 State space diagram for a system failing safely or unsafely either due to hardware failure or user error.

shown in Figure 13.4 [5]:

$$\frac{dP_0(t)}{dt} + (\lambda_s + \lambda_u)P_0(t) = \mu_s P_1(t) + \mu_u P_2(t) \tag{13.28}$$

$$\frac{dP_1(t)}{dt} + \mu_s P_1(t) = \lambda_s P_0(t) \tag{13.29}$$

$$\frac{dP_2(t)}{dt} + \mu_u P_2(t) = \lambda_u P_0(t). \tag{13.30}$$

At time $t = 0$, $P_0(0) = 1$, $P_1(0) = P_2(0) = 0$.
By solving Equations 13.28 through 13.30, we get

$$P_0(t) = \frac{\mu_s \mu_u}{y_1 y_2} + \left[\frac{(y_1 + \mu_s)(y_1 + \mu_u)}{y_1(y_1 - y_2)} \right] e^{y_1 t} - \left[\frac{(y_2 + \mu_s)(y_2 + \mu_u)}{y_2(y_1 - y_2)} \right] e^{y_2 t}, \tag{13.31}$$

where

$$y_1, y_2 = \frac{-C \pm [C^2 - 4(\mu_s \mu_u + \lambda_s \mu_u + \lambda_u \mu_u)]^{1/2}}{2}$$

$$C = \mu_s + \mu_u + \lambda_s + \lambda_u$$

$$y_1 y_2 = \mu_s \mu_u + \lambda_s \mu_u + \lambda_u \mu_s$$

$$y_1 + y_2 = -(\mu_s + \mu_u + \lambda_s + \lambda_u)$$

$$P_1(t) = \frac{\lambda_s \mu_u}{y_1 y_2} + \left[\frac{(\lambda_s y_1 + \lambda_s \mu_u)}{y_1(y_1 - y_2)} \right] e^{y_1 t} - \left[\frac{(\mu_u + y_2)\lambda_s}{y_2(y_1 - y_2)} \right] e^{y_2 t} \tag{13.32}$$

$$P_2(t) = \frac{\lambda_u \mu_s}{y_1 y_2} + \left[\frac{(\lambda_u y_1 + \lambda_u \mu_s)}{y_1(y_1 - y_2)} \right] e^{y_1 t} - \left[\frac{(\mu_s + y_2)\lambda_u}{y_2(y_1 - y_2)} \right] e^{y_2 t}. \tag{13.33}$$

Note that Equations 13.32 and 13.33 give the probability of the system failing safely and unsafely, respectively, when subjected to the repair process.

As time t becomes very large, the system steady-state probability of failing safely using Equation 13.32 is

$$P_1 = \lim_{t \to \infty} P_1(t) = \frac{\lambda_s \mu_u}{y_1 y_2}. \tag{13.34}$$

Similarly, as time t becomes very large, the system steady-state probability of failing unsafely using Equation 13.33 is

$$P_2 = \frac{\lambda_u \mu_s}{y_1 y_2}. \qquad (13.35)$$

Finally, as time t becomes very large, the system steady-state availability using Equation 13.31 is

$$AV = P_0 = \frac{\mu_s \mu_u}{y_1 y_2}, \qquad (13.36)$$

where AV is the system steady-state availability and P_0 is the system steady-state probability of operating normally.

EXAMPLE 13.4

Assume that an engineering system can fail safely or unsafely due to a hardware failure or a user error, and its constant safe and unsafe failure rates are 0.007 failures/h and 0.001 failures/h, respectively. Similarly, its safe and unsafe failure mode constant repair rates are 0.04 repairs/h and 0.002 repairs/h, respectively.

Calculate the probability of the engineering system being in unsafe failure mode during a very long mission period.

By substituting the given data values into Equation 13.35, we get

$$P_2 = \frac{\lambda_u \mu_s}{y_1 y_2} = \frac{(0.001)(0.04)}{(0.04)(0.002) + (0.007)(0.002) + (0.001)(0.04)}$$

$$= 0.2985.$$

Thus, the probability of the engineering system being in unsafe failure mode state during a very large mission period is 0.2985.

13.6 Model V

Model V represents an engineering system that can be in any one of three states: operating normally, operating unsafely due to a user error, and failed. The system is repaired from a fully failed state and unsafely operating due to a user error state. The system state space diagram is shown in Figure 13.5.

The following assumptions are associated with this model:

- The system failure and repair rates are constant.
- The repaired system is as good as new.
- System failures occur independently.

The following symbols are associated with the state space diagram shown in Figure 13.5 and its associated equations:

$P_j(t)$ is the probability that the system is in state j at time t, for $j = 0,1,2$.
λ_1 is the system constant failure rate from state 0 to state 2.
λ_2 is the system constant failure rate from state 0 to state 1.
λ_3 is the system constant failure rate from state 1 to state 2.
μ_1 is the system constant repair rate from state 2 to state 0.
μ_2 is the system constant repair rate from state 1 to state 0.
μ_3 is the system constant repair rate from state 2 to state 1.

With the aid of the Markov method described in Chapter 4, we write the following set of differential equations for the state space diagram shown in Figure 13.5 [5–7]:

$$\frac{dP_0(t)}{dt} + (\lambda_1 + \lambda_2)P_0(t) = \mu_1 P_2(t) + \mu_2 P_1(t) \tag{13.37}$$

$$\frac{dP_1(t)}{dt} + (\mu_2 + \mu_3)P_1(t) = \mu_3 P_2(t) + \lambda_2 P_0(t) \tag{13.38}$$

$$\frac{dP_2(t)}{dt} + (\mu_1 + \mu_3)P_2(t) = \lambda_3 P_1(t) + \lambda_1 P_0(t). \tag{13.39}$$

At time $t = 0$, $P_0(0) = 1$, and $P_1(0) = P_2(0) = 0$.

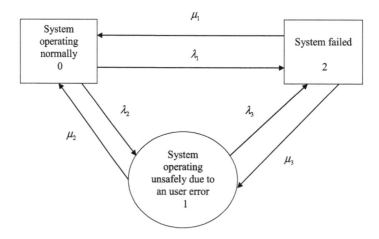

Figure 13.5 Engineering system state space diagram.

For a very large t, by solving Equations 13.37 through 13.39, we obtain the following steady-state probability equations:

$$P_0 = \frac{(\mu_1 + \mu_3)(\mu_2 + \lambda_3) - \lambda_3\mu_3}{A}, \tag{13.40}$$

where

$$A = (\mu_1 + \mu_3)(\mu_2 + \lambda_2 + \lambda_3) + \lambda_1(\mu_2 + \lambda_3) + \lambda_1\mu_3 + \lambda_2\lambda_3 - \lambda_3\mu_3$$

$$P_1 = \frac{\lambda_2(\mu_1 + \mu_3) + \lambda_1\mu_3}{A} \tag{13.41}$$

$$P_2 = \frac{\lambda_1\lambda_3 + \lambda_1(\mu_2 + \lambda_3)}{A}, \tag{13.42}$$

where P_0, P_1, and P_2 are the steady-state probabilities of the engineering system being in states 0, 1, and 2, respectively.

Note that the steady-state probability of the engineering system operating unsafely due to a user error is given by Equation 13.40.

By setting $\mu_1 = \mu_3 = 0$ in Equations 13.37 through 13.39, we obtain the following equation for the system's reliability (i.e., the probability of the system operating normally and unsafely):

$$R_s(t) = P_0(t) + P_1(t)$$
$$= (A_1 + B_1)e^{y_1 t} + [(A_2 + B_2)e^{y_2 t}], \tag{13.43}$$

where

$R_s(t)$ is the system reliability at time t.

$$y_1 = -M_1 + \left(M_1^2 - 4M_2\right)^{1/2}$$

$$y_2 = -M_1 - \left(M_1^2 - 4M_2\right)^{1/2}$$

$$M_1 = \mu_2 + \lambda_1 + \lambda_2 + \lambda_3$$

$$M_2 = \lambda_1\mu_2 + \lambda_1\lambda_3 + \lambda_2\lambda_3$$

$$A_1 = \frac{y_1 + \mu_2 + \lambda_3}{(y_1 - y_2)}$$

$$A_2 = \frac{y_2 + \mu_2 + \lambda_3}{(y_2 - y_2)}$$

$$B_1 = \frac{\lambda_2}{(y_1 - y_2)}$$

$$B_2 = \frac{\lambda_2}{(y_2 - y_1)}.$$

By integrating Equation 13.43 over the time interval $[0, \infty]$, we get the following equation for the system's MTTF:

$$MTTF_s = \int_0^\infty R_s(t)dt$$

$$= -\left[\frac{(A_1 + B_1)}{y_1} + \frac{(A_2 + B_2)}{y_2}\right],$$

(13.44)

where $MTTF_s$ is the system's MTTF.

EXAMPLE 13.5

Assume that an engineering system being used in the field can be either operating normally, operating unsafely due to a user error, or failed. Its failure rates from normal operating state to unsafe operating state, unsafe operating state to fully failed state, and normal operating state to fully failed state are 0.005, 0.002, and 0.08 failures/h, respectively. Similarly, the engineering system repair rates from fully failed state to normal operating state, unsafe operating state to normal operating state, and fully failed state to unsafe operating state are 0.06, 0.008, and 0.002 repairs/h, respectively.

Calculate the probability of the engineering system being in an unsafe operating state during a very large mission period, if the failure and repair rates associated with the system are constant.

By substituting the specified data values into Equation 13.41, we obtain

$$P_1 = \frac{(0.005)(0.06 + 0.002) + (0.08)(0.002)}{A}$$

$$A = (0.06 + 0.002)(0.008 + 0.005 + 0.002) + (0.08)(0.008 + 0.002)$$
$$+ (0.08)(0.002) + (0.005)(0.002) - (0.002)(0.002)$$
$$= 0.2478.$$

Thus, the probability of the engineering system being in an unsafe operating state during a very large mission period is 0.2478.

13.7 Model VI

Model VI is concerned with estimating the value of the glare constant. The model defines the glare constant as follows [8,9]:

$$G_c = \frac{\gamma_s^x L_s^y}{\gamma_v^2 L_{gb}},$$

(13.45)

where G_c is the glare constant, L_s is the source luminance, L_{gb} is the luminance of the general background, γ_s is the solid angle substanded by the source at the eye, γ_v is the angle between the viewing direction and the glare source direction, and x and y are the constants with given values of 0.8 and 1.6, respectively.

Note that $G_c = 35$ is considered as the boundary of "just acceptable" glare, and $G_c = 150$ is the boundary of "just uncomfortable" glare.

13.8 Model VII

Model VII is concerned with estimating movement time and is also known as the Fitts' Law model. With regard to usability, the model permits the designer to estimate the amount of time required for positioning, something like a computer cursor on a displayed object or item as a function of the distance to be traveled and the size of the item or object to be selected [10]. The movement time is expressed as follows [8,11]:

$$MT = \lambda + \gamma\mu,$$

(13.46)

where MT is the moment time, λ is the intercept constant, γ is the slope constant, and μ is the index of difficulty.

In turn, μ is expressed by

$$\mu = \log_2\left[\frac{2DSC}{WT}\right],$$

(13.47)

where DSC is the distance from the starting point to the center of the target and WT is the width of the target.

Note that Reference 12 proposed the following equation to estimate μ to correct for the tendency of Equation 13.46 to predict lower than observed movement times at rather low values of μ:

$$\mu = \log_2\left[\frac{DSC}{(WT + 0.5)}\right].$$

(13.48)

13.9   Model VIII

Model VIII is concerned with estimating the value of the *control-display (C/D) ratio*, which is the ratio of the control movement distance to that of the display moving element. Although this ratio is applicable only to continuous controls, it is considered to be a very important design factor that directly or indirectly affects operator performance. Some studies conducted over the years clearly indicate that a good C/D ratio can save from 0.5 to 5 s in positioning time compared to a poor one.

A C/D ratio for controls that involve considerable rotational movement and that affect linear displays is expressed as follows [9,13]:

$$\left(\frac{C}{D}\right) = \frac{X}{D_m} \tag{13.49}$$

$$X = \frac{2\pi(L_a)(C_{av})}{360}, \tag{13.50}$$

where D_m is the display movement, L_a is the lever arm length, and C_{av} is the control angular movement expressed in degrees.

Finally, note that the range of optimal values of the C/D ratio for ball levers and knobs, respectively, are as follows:

- 2.5:1 to 4:1
- 0.2 to 0.8

13.10   Model IX

Model IX is concerned with estimating the value of illuminance. It defines illuminance as follows [9,14]:

$$ILLU = \frac{L_i}{DLS}, \tag{13.51}$$

where ILLU is the illuminance/illumination expressed in lux (1 lumen/m^2), L_i is the luminous intensity (expressed in candelas), and DLS is the distance from the light source (expressed in meters).

13.11   Model X

Model X is concerned with estimating optimal character height. It uses the following equation to estimate the optimum character height [9,14,15]:

$$OCH = \frac{(RCH)(RD)}{(CSV)} \tag{13.52}$$

where RD is the required distance (expressed in inches), OCH is the estimated optimum height of a character at RD (expressed in inches), RCH is the recommended/standard character height at a viewing distance of 28 in., and CSV is a constant, with a specified value of 28 in. Note that the specific reason for setting CSV equal to 28 in. is that, for a comfortable arm reach for carrying out control and adjustment-related tasks, usually the instrument panels are located at a viewing distance of 28 in.

EXAMPLE 13.6

Assume that it is estimated that a meter of a newly installed system has to be read from a distance of 40 in. The recommended numeral height at a viewing distance of 28 in. and a low luminance is 0.38 in. Calculate the numeral height for the 40-in. viewing distance.

By substituting the specified data values into Equation 13.52, we obtain

$$OCH = \frac{(0.38)(40)}{(28)}$$

$$= 0.54 \text{ in.}$$

Thus, the numeral height for the 40-in. viewing distance is 0.54 in.

13.12   Model XI

Model XI is concerned with estimating brightness contrast, defined as follows [9,16]:

$$C_b = \frac{(L_b - L_d)(100)}{L_b}, \tag{13.53}$$

where C_b is the brightness contrast, L_b is the luminance of the brighter of two contrasting areas, and L_d is the luminance of the darker of two contrasting areas.

EXAMPLE 13.7

A certain type of paper has a reflectance of around 95%, and the reflectance of print on the paper is about 20%. Calculate the brightness contrast with the aid of Equation 13.53

By substituting the given data values into Equation 13.53, we obtain

$$C_b = \frac{(95 - 20)(100)}{(95)}$$

$$\cong 79\%.$$

Thus, the brightness contrast is about 79%.

13.13   Model XII

Model XII is concerned with estimating the required rest period for humans as they perform various types of tasks in their day-to-day work activity. The length of the rest period needed may vary significantly from one type of task to another. Thus, during the design of engineering systems for use by humans, the necessary user rest period must be taken into consideration for their ultimate effectiveness.

The model uses the following equation for estimating the required duration of a rest period for a specified task [9,17]:

$$R_p = \frac{WT(E_a - E_s)}{(E_a - \theta)}, \tag{13.54}$$

where R_p is the required rest period (expressed in minutes); WT is the total working time (expressed in minutes); E_s is the standard energy expenditure (expressed in kcal/min), whose value may be taken as 5 kcal/min when no data values are available; E_a is the average energy expenditure per minute of work (expressed in kcal/min); and θ is the resting level, whose approximate value is taken as 1.5 kcal/min.

EXAMPLE 13.8

Assume that a person is carrying out a computer-interactive task for 120 min, and his or her average energy expenditure is about 5.5 kcal/min. Calculate the length of the required rest period if the standard energy expenditure is 4 kcal/min.

By inserting the given data values into Equation 13.54, we obtain

$$R_p = \frac{(120)(5.5 - 4)}{(5.5 - 1.5)}$$

$$= 45 \text{ min}$$

Thus, the length of the required rest period is 45 min.

13.14   Model XIII

Model XIII is concerned with estimating the maximum lifting load for a person. It uses the following equation to estimate the maximum lifting load [9,18]:

$$MLL = (BMS_i)\alpha, \tag{13.55}$$

where MLL is the maximum lifting load, BMS_i is the isometric back muscle strength, and α is a constant whose values for females and males are 0.95 and 1.1, respectively.

13.15   Model XIV

Model XIV is concerned with estimating the loudness of sound. It uses the following equation to estimate the sound loudness [9,10]:

$$LS = 2^m \tag{13.56}$$

$$m = \frac{(SLL - 40)}{10}, \tag{13.57}$$

where LS is the loudness of a sound (expressed in sones) and SLL is the loudness level of a sound (expressed in phons).

13.16   Model XV

Model XV is concerned with measuring inspector performance associated with inspection-related tasks. It defines the inspector/human performance as follows [9,15]:

$$I_p = \frac{IT_t}{(\alpha - \theta)}, \tag{13.58}$$

where I_p is the inspector performance (expressed in minutes) per correct inspection, α is the total number of patterns inspected, θ is the total number of inspector errors, and IT_t is the total inspection time.

PROBLEMS

1. Assume that an engineering system can fail due to either a hardware failure or a user error, and its constant hardware failure and user error rates are 0.007 failures/h and 0.002 errors/h, respectively. Calculate the system failure probability due to user error for a 250-h mission.
2. Prove Equation 13.18 by using Equations 13.13 and 13.15.
3. Assume that the constant failure rate of a system is 0.004 failures/h. Calculate the system MTTF and its reliability during a 7-h mission.
4. Assume that an engineering system can fail safely or unsafely due to a hardware failure or a user error, and its constant safe and unsafe failure rates are 0.008 failures/h and 0.002 failures/h, respectively. Similarly, its safe and unsafe failure mode constant repair rates are 0.05 repairs/h and 0.001 repairs/h, respectively. Calculate the probability of the engineering system being in unsafe failure mode during a very large mission period.

5. Prove Equations 13.40 through 13.42 by using Equations 13.37 through 13.39.
6. Define glare constant.
7. What is the C/D ratio?
8. Assume that it is estimated that a meter of a newly installed system has to be read from a distance of 50 in. The recommended numeral height at a viewing distance of 28 in. at a low luminance is 0.38 in. Calculate the numeral height for the 45-in. viewing distance.
9. Assume that a certain type of paper has a reflectance of 90% and the reflectance of print on the paper is 18%. Calculate the value of the brightness contrast.
10. Assume that a person is performing a computer-interactive task for 140 min and his/her average energy expenditure is 5 kcal/min. Calculate the length of the required rest period if the standard energy expenditure is 5 kcal/min.

References

1. Dhillon, B. S., *Mine Safety: A Modern Approach*, Springer-Verlag, London, 2010.
2. Dhillon, B.S., *Safety and Reliability in the Oil and Gas Industry: A Practical Approach*, CRC Press, Boca Raton, FL, 2016.
3. Dhillon, B. S., Stochastic Models for Predicting Human Reliability, *Microelectronics and Reliability*, Vol. 25, 1985, pp. 729–752.
4. Dhillon, B. S., and Rayapati, S. N., Reliability and Availability Analysis of Transit Systems, *Microelectronics and Reliability*, Vol. 25, No. 6, 1985, pp. 1073–1085.
5. Dhillon, B. S., *Engineering Systems Reliability, Safety, and Maintenance: An Integrated Approach*, CRC Press, Boca Raton, FL, 2017.
6. Dhillon, B. S., *Engineering Safety: Fundamentals, Techniques, and Applications*, World Scientific Publishing, River Ridge, NJ, 2003.
7. Dhillon, B. S., and Kirmizi, F., Probabilistic Safety Analysis of Maintainable Systems, *International Journal of Quality in Maintenance Engineering*, Vol. 9, No. 3, 2003, pp. 303–320.
8. Fitts, P. M., and Peterson, J. R., Information Capacity of Discrete Motor Responses, *Journal of Experimental Psychology*, Vol. 67, 1964, pp. 103–112.
9. Dhillon, B. S., *Engineering Usability: Fundamentals, Applications, Human Factors, and Human Error*, American Scientific Publishers, Stevenson Ranch, CA, 2004.
10. McMillan, G. R., Eggleston, R. G., and Anderson, T. R., Non-Conventional Controls, in *Handbook Human Factors and Ergonomics*, edited by G. Salvendy, John Wiley & Sons, New York, 1997, pp. 729–827.
11. Fitts, P. M., The Information Capacity of the Human Motor System in Controlling the Amplitude of Movement, *Journal of Experimental Psychology*, Vol. 47, 1954, pp. 381–391.

12. Welford, A. T., The Measurement of Sensory-Motor Performance: Survey and Reappraisal of Twelve years' Progress, *Ergonomics*, Vol. 3, 1960, pp. 189–230.
13. *Human Engineering Guide to Equipment Design*, Sponsored by Joint Army-Navy-Air Force Committee, John Wiley Sons, New York, 1972.
14. McCormick, E. J., and Sanders, M.S., *Human Factors in Engineering Design*, McGraw Hill Book Company, New York, 1982.
15. Drury, C. G., and Fox, J.G., editors: *Human Reliability in Quality Control*, John Wiley & Sons, New York, 1975.
16. Dhillon, B. S., *Human Reliability: With Human Factors*, Pergamon Press, New York, 1986.
17. Murrell, K. F. H., *Human Performance in Industry*, Reinhold Publishing Company, New York, 1965.
18. Poulsen, E., and Jorgensen, K., Back Muscle Strength, Lifting and Stooped Working Postures, *Applied Ergonomics*, Vol. 2, 1971, pp. 133–137.

Index